Active Electronically Scanned Arrays

IEEE Press
445 Hoes Lane
Piscataway, NJ 08854

Active Electronically Scanned Arrays

Fundamentals and Applications

Arik D. Brown, PhD

IEEE PRESS
WILEY

Published by John Wiley & Sons, Inc., Hoboken, New Jersey.
Published simultaneously in Canada.

Library of Congress Cataloging-in-Publication Data is applied for
Hardback ISBN: 9781119749059

Cover Design: Wiley
Cover Image: © Graphic by Arik Brown

Set in 9.5/12.5pt STIXTwoText by Straive, Chennai, India

10 9 8 7 6 5 4 3 2 1

FSC
www.fsc.org
MIX
Paper from responsible sources
FSC® C013604

To God, my wife Nadine, my children Alexis and Joshua, my parents Rudy and Meredith, my ever-supportive family, and in particular, my grandfather, the late James Gibby Sr., whose love for education lives on through me. Love has motivated me, faith has activated me, the power of the Spirit has accelerated me, and wisdom has provided the navigation. Thanks, Pastor Mike.

Contents

Preface

For those unfamiliar with AESAs, they can seem very confusing and complicated. However, there are basic fundamental principles that drive how AESAs operate and what are key metrics for their performance. This book's goal is to provide the AESA beginner, whether a student, researcher, or practicing engineer, an understanding of why AESAs are critical technology, what are the primary subsystems in an AESA, key driving subsystem parameters that govern AESA performance, and various AESA topologies that are commonly implemented.

AESAs provide incredible capability working synergistically with downconverters, upconverters, exciters, and/or receivers. Having worked with AESAs for over 20 years, I have grown accustomed to thinking of the entire system from the AESA front-end elements to the backend electronics as the "AESA", but in practice the AESA consists of the front-end array elements and the electronics used to provide signal conditioning, amplification, and beam steering. One of my colleagues (John Welch) and I had a friendly disagreement on what defined an AESA, and he'll be amused to know that I now agree with his definition. This discussion is relevant to this book as its chapters are organized in this manner.

Chapter 1 provides a history of AESA development going back to the 1960s. Differences between mechanically scanned arrays (MSAs), passive electronic scanned arrays (PESAs) and AESAs are discussed and elaborated upon. Also included is a summary of the benefits that AESAs provide for different applications. In any real system, the end user cares most about how his or her mission will be accomplished and is not mesmerized by the fact that the system utilizes AESA technology. Finally, Chapter 1 concludes with a system block diagram of an AESA that will be used as a point of reference throughout this book. Chapters 2 through 5 cover AESA theory and the primary subassemblies that comprise an AESA: array elements, transmit receive modules (trms), and beamformer. Chapter 2 describes theoretically how beams are electronically scanned and how AESA performance can be modeled. Further details on AESA modeling with Matlab code examples can be found in [Brown, 2012]. Chapter 3 focuses on the array elements.

Fundamental aspects of performance such as element spacing, grating lobes, active impedance, and scan loss are discussed. Chapter 4 discusses transmit receive modules. TRMs provide signal conditioning (i.e., RF filtering), amplification, and phase and/or time delay for steering. Chapter 5 focuses on the the beamformer. In practice, the beamformer can also be referred to as the RF manifold. This chapter will show how signals are distributed from the exciter and also how signals received from the array elements are combined together to form beams.

The final two chapters cover AESA cascaded performance and architectures that are common in AESA systems. Chapter 6 describes how to calculate cascaded parameters such as signal and noise gain, noise factor, and intercept point for an AESA. Using these fundamental principles any AESA can be represented by its cascaded response. Additionally, using the cascaded intercept point and noise factor, an expression for the AESA spurious free dynamic range (SFDR) is defined. Chapter 7 provides patterns for various AESA topologies spanning AESAs with phase shifters, AESAs with true time delay, subarrayed AESAs, and elemental digitally beamformed AESAs. It concludes with an overview on adaptive beamforming and several pattern examples.

I have worked on various systems spanning radar, electronic warfare, communications, and SIGINT for over 20 years. I have found that no matter what the application, the fundamentals remain "the fundamentals". My hope is that these fundamentals are brought to light and made tangible to the reader uninitiated to AESA technology. Also, my goal is that those who are or have worked on AESAs will find utility in this book for explaining topics that in a normal busy day they have no time to research or investigate. As a system architect, I have had to take system level requirements and derive lower-level requirements that directly affect an AESA. I bring some of that flavor in the following chapters.

Maryland
March, 2021

Arik D. Brown

Acknowledgments

To the many great people who have imparted wisdom to me on my journey, and especially Dr. John Volakis, Dr. Leo Kempel, Uncle Bill (Bill Hopwood), Lee Cole, Carl Nothnick, John Wotjowitcz Sr., Rich Konapelsky, Dr. Daniel Boeringer, Dr. Eric Holzman, Dan Davis, Steve Pettit, Clint Ung, and Bob Worley.

Arik D. Brown

Acronyms

ABF	adaptive beamforming
AESA	active electronically scanned array
AoA	angle of arrival
AUT	antenna under test
AR	axial ratio
BIT	built-in test
CCW	counterclockwise
CPI	coherent processing interval
CW	clockwise
dB	decibels
DBF	digital beamforming
eCHR	enhanced compact hemispheric radar
EA	electronic attack
EDBF	elemental digital beamforming
ERP	effective radiated power
ESM	electronic support measures
EW	electronic warfare
FOV	field of view
GaAs	gallium arsenide
GaN	gallium nitride
HPA	high power amplifier
ieMHR	improved and enhanced multi-mission hemispheric radar
IBW	instantaneous bandwidth
IP2	second-order intercept point
IP3	third-order intercept point
LHCP	left-hand circular polarization
LHEP	left-hand elliptical polarization
LNA	low noise amplifier
MSA	mechanically scanned array

MTBF	mean time between failures
PAE	power added efficiency
PESA	passive electronically scanned array
RADAR	radio detection and ranging
RHCP	right-hand circular polarization
RHEP	right-hand elliptical polarization
RRE	radar range equation
SAR	synthetic aperture radar
SFDR	spurious free dynamic range
SL	sidelobe
SLL	sidelobe level
SNR	signal-to-noise ratio
TR	transmit receive
TRM	transmit receive module

1

AESA Overview

1.1 Introduction

The past 30 years have seen a significant increase in the capability of active electronically scanned arrays (AESAs). What was once deemed a relatively new and expensive technology has now proliferated in many applications such as defense, communications, and the automotive industry. The advancement of microelectronics and receiver technology has also benefited AESAs. The ability to produce microelectronic circuits and high-speed wide bandwidth receivers in a small form factor has made AESAs a realizable commodity for array applications. Additionally, and arguably most crucial, is the ability to produce AESAs cost effectively. In their infancy stage, AESAs were deemed cost prohibitive for any application outside of the defense industry (specifically airborne radar). This, however, is no longer the case. As an example, the automotive industry uses small inexpensive AESA radars for collision avoidance (Tokoro et al., 2003).

1.2 AESA History

Although AESAs are becoming more widespread, many people unfamiliar with AESAs believe it is a new technology. However, AESAs were being developed as far back as the early 1960s. The driving application was radar where agile and simultaneous beams are highly advantageous for tracking fast-moving targets. In 1960, Bell Labs proposed a phased array to replace the Nike Zeus radar (Figure 1.1) (Labs, 1975). This radar used a distributed phased array approach employing multiple reflector antennas that were spaced apart. Each reflector antenna had its own transmitter and receiver. This enabled simultaneous beams accomplishing detection and tracking at the same time. Key features of this radar were long distance detection, track generation, discrimination of warheads from decoys, and tracking of outbound interceptor missiles (Labs, 1975).

Active Electronically Scanned Arrays: Fundamentals and Applications, First Edition. Arik D. Brown.

Figure 1.1 Zeus Multi-Function Array Radar, developed by Bell Labs, utilized a distributed phased array. (Labs, 1975)

In the early 1970s, airborne radar became the driver for AESA development. AESAs opened up new possibilities to provide superior air dominance. Agile beams that could be scanned rapidly on the order of microseconds with no moving parts (gimbal), higher reliability (MTBF) and graceful degradation in performance were some of the key features that caused a flurry of development to build high-performance airborne AESA radars. Over time this led to the development of AESAs that supported multi-function capability such as search, track while scan, synthetic aperture radar, and precision geolocation.

The 1980s and 1990s saw the expansion of AESAs from air to maritime and land. The increase in the repeatability of microelectronic circuits and the decrease in the cost of development made AESAs tenable for large arrays that could be used for maritime- and land-based radar systems. An airborne radar is limited in how big it can be for fighter applications and to some extent for long-range surveillance, but maritime and land radars do not have the same power and weight limitations. In order for a very large AESA to be built (e.g., > 10s of square meters), cost cannot be prohibitive.

In the 2000s there was an increase in the development of AESAs for Ka Band. AESAs at these frequencies can leverage semiconductor wafer processing providing a dramatic decrease in cost. AESAs at these frequencies can support a large number of elements (>> 1,000) and also enable the ability to use wider bandwidths due to the increase in frequency. As an example, for an AESA built for X band, a 22 percent bandwidth represents $\sim$ 2 GHz of operational bandwidth;

however, at Ka band a 22 percent bandwidth represents ~ 8 GHz of operational bandwidth. For applications such as communications, this is very attractive due to the increased spectrum for the same percent bandwidth. In addition to communications, the automotive industry has leveraged AESAs at Ka band frequencies for radars that support automotive collision prevention.

From approximately 2010 until the present, we are now seeing a growing increase in the use of AESAs by smaller-sized companies that are able to build more affordable AESAs for radar applications such as active protection systems (APS), counter unmanned aerial systems (C-UAS), counter rockets artillery and mortars (C-RAM) and short-range air defense (SHORAD). Examples of this are the enhanced compact hemispheric radar (eCHR) and improved and enhanced multi-mission hemispheric radar (ieMHR) (Figure 1.2) built by RADA. These 4D AESA radars have an excellent performance-to-price ratio, are software defined, and support multi-mission operation.

Currently AESAs are also extending beyond radar into applications such as electronic attack (EA), signals intelligence (SIGINT), and electronic support measures (ESM). These capabilities existed in airborne AESA radars; however, they were adjunct capabilities to the primary radar function. We're now seeing AESA systems that are are specifically designed for the applications previously mentioned. As an example, the use of AESAs for EA on the mid-band pods for the E-18G Growler (Reim, 2021) shown in Figure 1.3, in addition to InTop (Grumman, 2019), show how the use of AESAs has expanded beyond solely radar.

1.3 AESA Applications

As previously mentioned, the original force and impetus behind AESA development and application was radar. AESA technology advancements, especially in the area of airborne radar, created a technology base that could be used for other applications. The features and characteristics that make AESAs attractive for radar have a similar attraction for other technology areas. The applicability of AESAs for these areas is described next.

1.3.1 RADAR

Initial radars used mechanically scanned arrays (MSAs). MSAs use a mechanical gimbal to point a reflector antenna. An example is illustrated in Figure 1.4 that shows the MSA for the APG-78. An advantage of MSAs is that they do not suffer scan loss. (Scan loss will be explained in Chapter 3.) However, the speed with which the array beam can be scanned spatially is limited by the gimbal. AESAs

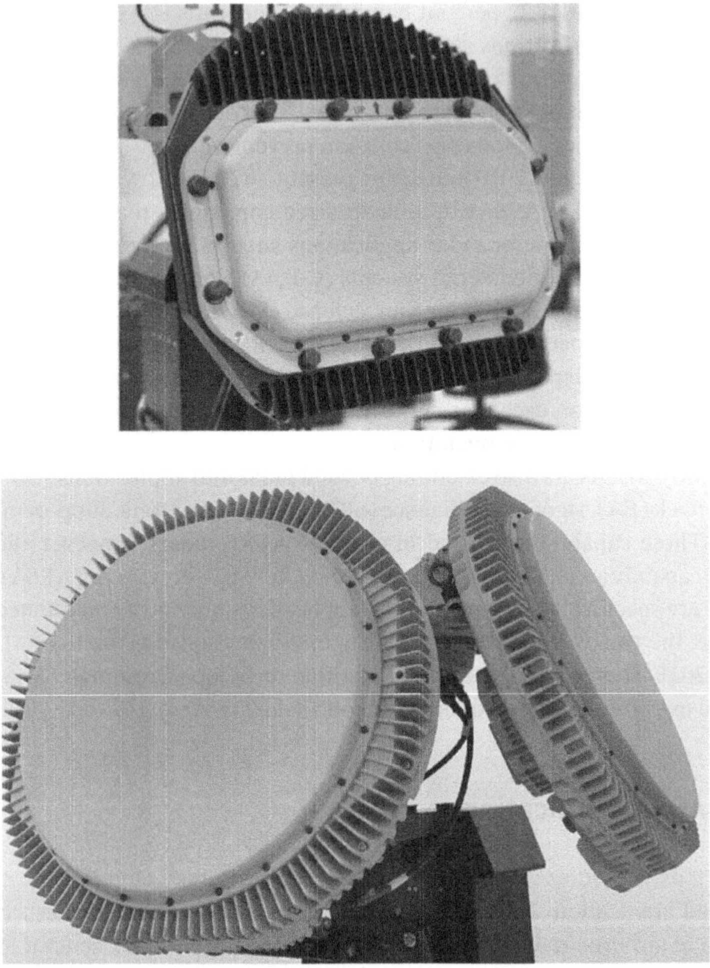

Figure 1.2 RADA's eCHR (top image) and ieMHR (bottom image) utilize AESAs to provide multi-mission capability.

enabled a substantial increase in scan volume speed on the order of milliseconds to microseconds. Current AESAs can scan beams on the order of nanoseconds. Figure 1.5 shows the difference in how beam scanning is accomplished between an MSA and an AESA. The AESA has phase shifter devices that apply a different phase at each array element. This causes the array beam to coherently add at any commanded scan angle, which is called beam scanning (Figure 1.5). The theory behind this will be covered in Chapter 2, and the details of a transmit receive module (TRM) will be explained in Chapter 4. TRMs contain the phase shifter devices that enable beam scanning. Since this scanning is done electronically, it is

Figure 1.3 The US Navy is employing AESA technology on the Next Generation Jammer mid-band pods for the EA-18G Growler. (US Navy photo by Erik Hildebrandt)

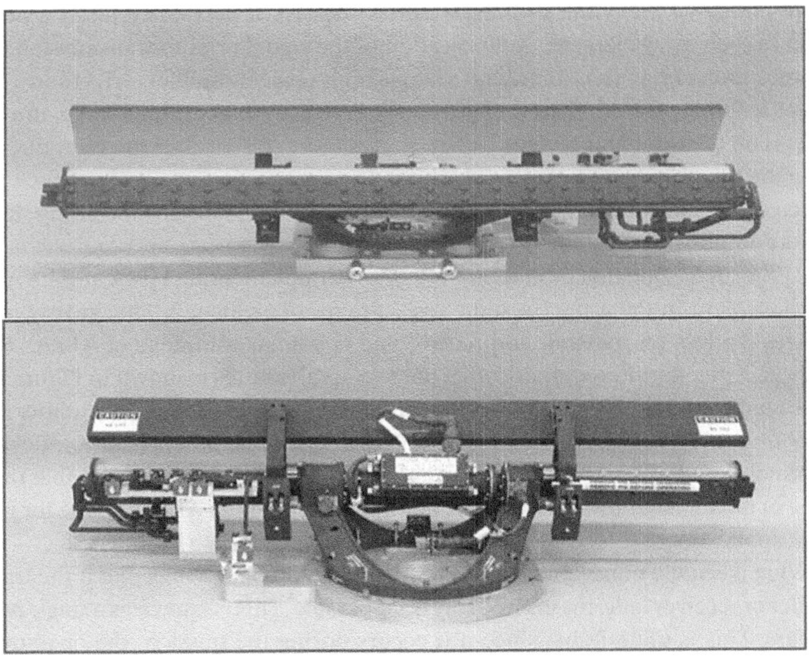

Figure 1.4 The APG-78 is an example of an early radar that used an MSA. Although effective, it does not have the beam agility of an AESA.

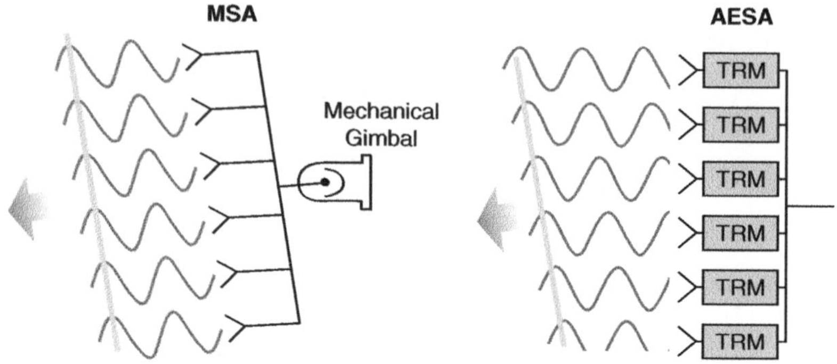

Figure 1.5 AESAs are able to electronically scan the array beam as opposed to mechanical scanning by MSAs. This provides orders of magnitude improvement in spatial scanning ability.

inherently faster than the mechanical instantiation of beam scanning by the MSA. This is commonly called *beam agility*.

The transition from MSAs to AESAs for radar was not a direct leap. A transitory architecture was developed that is called a passive electronically scanned array (PESA). This architecture is shown in Figure 1.6. The only electronics in the front-end array are phase shifters, which, as described previously, apply a phase shift to each array element. Amplification of the signal prior to transmission and on reception of signals is done with a single high-power amplifier (HPA) and a low noise amplifier (LNA). This architecture inherently suffers from a large amount of loss on transmit and receive. When transmitting, losses after the HPA directly degrade the transmitted power, which for a radar affects range performance. On receive, the same phenomenon occurs, and the loss attenuates the incoming signal before amplification by the LNA.

AESAs overcame this limitation of PESAs by implementing HPAs and LNAs at each element in the array (in some cases groups of elements). This greatly minimizes the loss on transmit and receive and is a large advantage of AESAs over PESAs. A side-by-side comparison of the two architectures is shown in Figure 1.6. The challenge initially for AESAs was to make the microelectronic circuits that contain the HPAs, LNAs, phase shifters, TR switches, and other electronic devices repeatable and affordable. GaAs was an important breakthrough to enable these type of microelectronic devices to be built, thereby making AESAs viable for mission applications.

What is readily noticeable in the PESA architecture diagram is that if the transmitter or receiver fails, the entire array is inoperable. This is known as a *single point failure*. This is undesirable since if it occurs during the mission, the operator is unable to use the radar, and also the transmitter and/or receiver must be replaced. Both of these issues have a negative effect on reliability and specifically MTBF. This is a critical performance parameter especially for DoD applications.

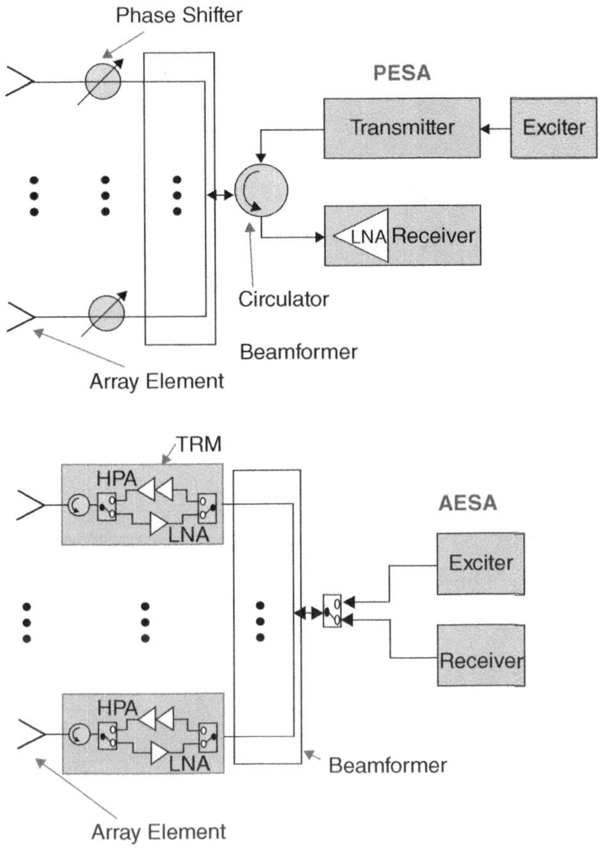

Figure 1.6 An AESA is a distributed architecture in contrast to a PESA. This provides advantages in increased radiated power and increased sensitivity on receive due to minimization of losses as compared to the centralized transmitter and receiver for the PESA.

Unlike PESAs, AESAs greatly improve MTBF. This is because if a single TRM fails, the array is still operational. In fact, depending on the size of the AESA (number of elements), as many as 6–10 percent of the elements could fail, and the array would still be operational and perform. This is commonly referred to as *graceful degradation*. This will be discussed in more detail in Chapter 2. Figures 1.7 and 1.8 show two examples of AESAs that exhibit graceful degradation.

1.3.2 Electronic Warfare

Airborne AESA radars have implemented EA and ESM modes for many years. However, the primary mode is typically radar related (search/surveillance,

Figure 1.7 The APG-81 AESA demonstrates graceful degradation due to the large number of array elements.

Figure 1.8 Similar to the APG-81, the APG-63 V2 AESA is another example of graceful degradation due to the large number of array elements. (Figure courtesy of Raytheon.)

tracking, SAR, etc.) while EA/ESM modes are ancillary. AESAs are now being used for EW functions only, which is a paradigm shift from years past. Examples of this are the Next Generation Jammer (Reim, 2021) and Integrated Topside (Grumman, 2019) programs for the Navy. Both are employing AESAs for EW use.

1.3.2.1 Electronic Attack

For EA, AESAs provide several key benefits to be used in an EA system. These benefits are described below:

- Directional high gain beams - Traditional EA systems use antennas that are omnidirectional or have very little gain. This means the effective radiated power (ERP), which is the product of the antenna gain and transmitted power, is primarily going to be driven by the available transmitted power since the antenna has a small amount of gain. AESAs are advantageous relative to ERP because they have large antenna gains and have the ability to transmit a large amount of power. This dramatically increases the standoff range for which the EA system can jam threats.
- Simultaneous beams - In typical EA engagements, there are multiple threats within the system's field of view (FOV) posing imminent danger to pilots flying in the vicinity. These threats can be spatially and frequency diverse. AESAs have the ability to be broken into subarrays allowing sections of the array to operate independently. Additionally, multiple beamformers can be implemented in the same AESA providing the ability to transmit beams at different frequencies by power sharing of the array. This makes AESAs optimal for defeating threats that are spatially and frequency diverse.
- Beam agility - Multiple threats can be jammed by either changing frequency for threats located in the same location or changing the frequency and location of the jamming beam for threats in different locations. The latter is commonly referred to as spatial commutation. Because AESAs have agile beams than can be scanned on the order of nanoseconds, they can be employed for spatial commutation.
- Wide instantaneous bandwidth (IBW) - Threats have evolved by employing wider bandwidth lower-power waveforms to search and detect. To counter this the jamming system has to be able to mimic these waveforms to reproduce and re-transmit for jamming. AESAs now have the capability to support large IBWs, which can be used in these situations.

1.3.2.2 Electronic Support Measures

Benefits for ESM are described below. The features an AESA provides for ESM are similar to EA, but the benefits are different because ESM employs passive reception of threat signals. AESAs that can support *both* EA and ESM are extremely powerful.

- High gain beams - Receive sensitivity is of utmost importance for ESM. It determines the smallest amplitude threat signal that can be detected for geolocation or other purposes. AESAs have high gain beams, which translates to superior sensitivity.
- Simultaneous beams - High gain beams have much smaller beamwidths than traditional omni-directional ESM antennas. This means the ability to search a given volume in space will take longer. To counter this, simultaneous beams provide the ability to search a spatial volume faster *and* with high sensitivity.
- Beam agility - Similar to simultaneous beams, beam agility enables the AESA to scan a spatial volume rapidly.
- Wide IBW - This also enables the ability to scan faster in the *frequency domain*. Being able to interrogate the frequency spectrum of interest rapidly is critical. AESAs can support this by using true time delay (discussed in Chapter 2).

1.3.3 Communications

With the expansion of many communication systems to Ka band and higher, AESAs are now highly attractive for communication applications. For low-frequency communications, AESAs are not a suitable choice since they would have to be extremely large and in most cases cost prohibitive. For higher frequencies, this barrier is removed, and AESAs can be used. The benefits for communications are described below.

- High gain beams - Similar to EA, high gain beams increase the range of communication systems. However, of perhaps even greater importance is that high gain beams increase SNR, which allows the use of more complex waveforms for increased data transmission rate.
- Simultaneous beams - This feature of AESAs allows a single system to reach multiple users at the same time. This is highly valuable for creating a persistent and reliable communications system.
- Beam agility - The beam agility of AESAs enables them to service users in diverse spatial locations. This eliminates the need of using multiple antennas to cover different regions spatially.
- Wide IBW - Wider bandwidths enables increased data throughput. It also supports more efficient use of the transmission frequency spectrum for waveforms such as OFDM. AESAs enable the ability to transmit and receive high gain, and simultaneous, agile beams with a wide IBW.

1.3.4 Signals Intelligence

SIGINT can be thought of as requiring only an ESM system. This means that the SIGINT AESA does not transmit and only receives signals from the environment.

Its sole purpose is to find small and large signals[1] across frequency with the ability to geolocate the signals and/or demodulate them for intelligence. The AESA benefits described for ESM are directly applicable for SIGINT. An interesting point to highlight is that since SIGINT AESAs do not transmit, they then have more space for additional receive electronics. This opens up the design space for increasing the number of beams in the AESA, making it even more powerful for rapidly scanning in both the spatial and frequency domains.

1.4 AESA Point of Reference

Before introducing the AESA block diagram, it is necessary to first review the relation between received and transmitted power from an array/antenna. This relation can be described with an equation that is extensively used in radar that is called the radar range equation (RRE), which is based upon the Friis transmission equation (Balanis, 1982). This equation represents the signal power that is transmitted, reflected from a target, and received by the same or another antenna, and the ratio of that power to the noise produced in the environment and by the system electronics, which is called the signal-to-noise ratio (SNR) (Balanis, 1982) and (Skolnik, 1990). For transmit-only and receive-only applications, this equation is still applicable and can be used by omitting certain parameters, which will be discussed in the following paragraphs. AESAs are a major contributor to the RRE for both the signal power and the noise power, so understanding the RRE is important for AESAs. The driving requirements flowed to an AESA are based upon system-level performance parameters that affect the RRE.

The RRE will first be described in the context of a radar, and later it will be pointed out how it applies for other non-radar applications. First, the signal power in the RRE will be addressed. Consider an antenna with a transmit gain G_{TX} and power P_{TX}. The product of P_{TX} and G_{TX} is the effective radiated power and represents the amount of power radiated in the far-field. The far-field region is a prescribed distance away from an antenna where the field intensity is essentially independent of the distance from the antenna (Balanis, 1982). This distance is calculated using the equation $R = \frac{2\,D^2}{\lambda}$, where R is the distance from the antenna, D is the maximum overall length of the antenna, and λ is the RF wavelength at the radiated frequency ($\lambda = \frac{c}{f}$, where c is the speed of light and f is the RF radiated frequency) (Balanis, 1982). In general, G_{TX} is a function of angle (θ, ϕ); however, for the purposes here it is assumed that G_{TX} represents the peak of the antenna gain

1 This is referred to as dynamic range and is calculated for an AESA in Chapter 6.

pattern. The radiated power decreases proportionally by the distance or range (R). The radiated power can then be represented as:

$$\frac{P_{TX}G_{TX}}{4\pi R^2}\left[\frac{\mathrm{W}}{\mathrm{m}^2}\right]. \tag{1.1}$$

The radiated power in Equation 1.1 is then incident upon an object or target, which reflects and effectively reradiates this energy. The parameter that describes this is the radar cross section, σ [m^2], and represents the amount of energy that is reflected back in the direction of the radiating antenna. Similar to the original radiated power the reflected power, decreases proportional to R. The power that arrives back at the radiating antenna can be represented by:

$$\frac{P_{TX}G_{TX}\sigma}{(4\pi)^2R^4}\left[\frac{\mathrm{W}}{\mathrm{m}^2}\right]. \tag{1.2}$$

The last step in computing the total reflected power at the transmitting antenna is to multiply the power in Equation 1.2 by the area of the antenna. The total received power is shown in Equation 1.3 and will be referred to as S. It will be shown that this S is the signal power for the SNR.

$$S = \frac{P_{TX}G_{TX}\sigma A}{(4\pi)^2R^4}\ [\mathrm{W}] \tag{1.3}$$

The A in Equation 1.3 is the effective capture area of the antenna and does not represent the actual physical dimension of the array. This equation form of S is the most generic in that it can be applied to both monostatic (transmitting and receiving antennas are the same) and bistatic (transmitting and receiving antennas are different and spatially separated) radars. For the the purposes of this discussion it is assumed monostatic. With this assumption, Equation 1.3 can modified by using the familiar expression of $G = \frac{4\pi A}{\lambda^2}$. Equation 1.3 can then be written as:

$$S = \frac{P_{TX}G_{TX}^2\sigma\lambda^2}{(4\pi)^3R^4}\ [\mathrm{W}]. \tag{1.4}$$

In Equation 1.4, it is seen that in order to maximize the signal power S for a given R, P_{TX} and G_{TX} must be maximized as well. As discussed earlier, AESAs inherently have high gain and are capable of transmitting large amounts of power. This is how the AESA directly affects and improves S. As an example, the effect of an MSA on S is calculated using the same expression in Equation 1.4. The difference is that reflector antennas are much less efficient than an AESA and have losses on the order of 60%. AESAs typically have losses on the order of 90%, which directly maximizes S in comparison to an MSA. Additionally, as was shown in Figure 1.6, the losses after the transmitter in an MSA are reduced with an AESA due to the AESA's distributed HPA architecture. This directly affects P_{TX} and thereby S.

To complete the SNR equation, the noise power N that competes with S must be calculated. The expression for N is shown in Equation 1.5. k [W/Hz*K] is

Boltzmann's constant, T_0 [K] is the noise temperature, B [Hz] is the noise bandwidth, and F is the noise factor, which represents the noise added by the system to the incoming noise from the environment (Pettai, 1984).

$$N = kT_0BF \text{ [W]}. \tag{1.5}$$

In an AESA, the LNAs are placed close to the array elements minimizing the added loss and therefore minimizing F. This decreases the amount of noise that is seen by the receiver and competes with S. The equation for SNR can then be written as:

$$SNR = \frac{S}{N} = \frac{\frac{P_{TX}G_{TX}^2\sigma\lambda^2}{(4\pi)^3R^4}}{kT_0BF}, \tag{1.6}$$

which can be simplified as:

$$SNR = \frac{P_{TX}G_{TX}^2\sigma\lambda^2}{(4\pi)^3R^4kT_0BF}. \tag{1.7}$$

From a design perspective, Equation 1.7 highlights the performance parameters that are affected by the AESA. These parameters are P_{TX}, G_{TX}, and F. For an optimal AESA design, the transmit power, array gain, and noise figure must be optimized and balanced to achieve a high-performance, cost-effective solution.

Thus far, the RRE equation has been discussed from a radar perspective of transmitting and receiving energy for system operation. Next, the modifications to the RRE will be discussed for application to EA and ESM. For EA, the AESA is used to transmit as much power as required to jam the target of interest. Because of this, the receive noise power is not a driver, and the primary consideration is the signal power, S, as previously shown in Equation 1.4. A modification has to be applied to S because for jamming there is only a one-way loss from the AESA to the target, and there is no radar cross section to be concerned with. Taking these factors into consideration, Equation 1.4 then becomes:

$$S_{EA} = \frac{P_{TX}G_{TX}}{(4\pi)R^2} \cdot A_{threat} \text{ [W]}. \tag{1.8}$$

As previously discussed, P_{TX} and G_{TX} are maximized to deliver as much power as possible and are key drivers for the AESA design. Also, unlike in the radar case, the signal power for EA only has to overcome an R^2 loss. Lastly, A_{threat}, is the effective area of the threat antenna and is not a parameter that is affected by the AESA.

For ESM, a modification to Equation 1.7, has to be done for both the signal power, S, and the noise power, N. In an ESM system, the AESA operates in a receive-only mode. For this case, there is no transmit power from the AESA, and the power received is from the threat or signal of interest. Also, similar to the EA case previously described there is only an R^2 loss to be considered from the signal of interest to the AESA. Equation 1.7 can then be modified as:

$$SNR_{ESM} = \frac{(P_{TX}G_{TX})_{threat}}{4\pi R^2} \cdot \frac{A_{AESA}}{kT_0BF}. \tag{1.9}$$

This can be further simplified using the familiar relation $A = \frac{\lambda^2 \cdot G}{4\pi}$:

$$SNR_{ESM} = (P_{TX}G_{TX})_{threat} \cdot \left(\frac{\lambda}{4\pi R}\right)^2 \cdot \frac{1}{kB} \cdot \frac{G}{T}. \tag{1.10}$$

From Equation 1.10, a key performance parameter is shown in the form of $\frac{G}{T}$, where T is equal to T_0F. $\frac{G}{T}$ is commonly referred to as *sensitivity* and directly affects how well the ESM system can detect signals in the presence of noise. A sensitivity requirement is typically flowed down to the AESA and drives AESA performance. A sensitivity requirement is also flowed down for radar and communication AESA systems. Figure 1.9 provides a graphical illustration of the AESA performance parameters that affect performance for radar, EA, and ESM. Communications, although not pictured, can be represented as a combination of EA and ESM in terms of functionality. From the AESA's perspective, the communications link must be closed (i.e. sufficient SNR) on transmit and receive, which is what is represented by the EA and ESM illustrations in Figure 1.9.

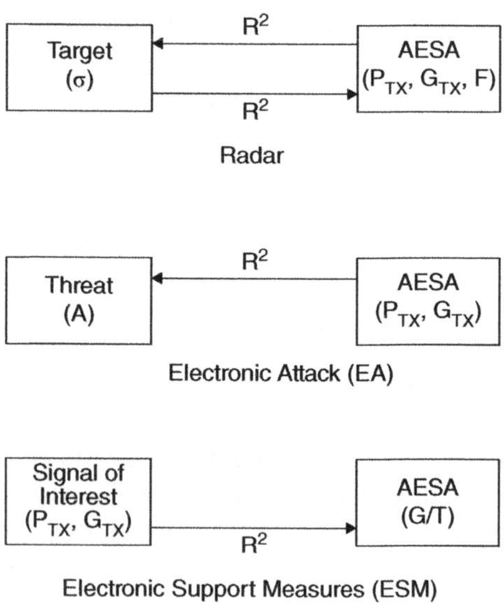

Figure 1.9 Illustration of the key performance parameters for an AESA that affect performance for radar, EA, and ESM.

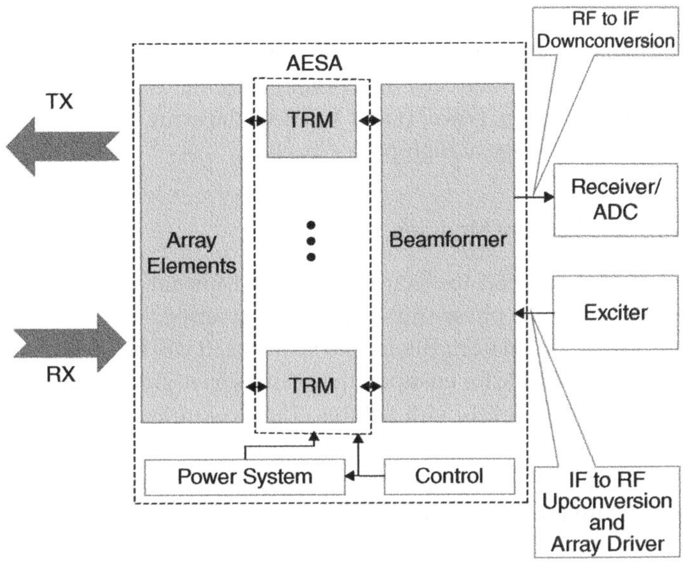

Figure 1.10 AESA block diagram highlights the major subsystems that comprise an AESA in addition to the backend electronics. These subsystems will be expounded upon in later chapters.

1.5 Block Diagram

With an understanding of how an AESA impacts SNR, the block diagram for a transmit receive (TR) system will be introduced that will be referenced throughout the remaining chapters. Figure 1.10 illustrates the relationship of how the AESA and backend electronics are connected. Effectively, the AESA can be simplified as an effective antenna with an effective HPA and LNA. This governs how the SNR will be affected. Traditionally in designing a system, the AESA will be flowed a requirement for total F, G, and P; this then is flowed down to the array elements, TRMs, and beamformer.

1.5.1 Antenna Array Elements

The antenna array elements are at the front end of the AESA. They directly affect the AESA antenna gain G in addition to polarization requirements. The array elements typically are either integrated with a radome or covered by an radome enclosure that is used to protect the elements from the environment. The radome is usually constructed to have minimal loss and also very small reflection for good transmittivity. Chapter 3 will provide a detailed explanation on the key array element performance parameters such as scan loss, active match, and insertion loss. The array elements also indirectly affect the requirement for P. For systems that transmit a large amount of power (radar and EA), the array elements must be able to operate without damage at high-power levels.

1.5.2 Transmit Receive Modules

The TRMs enable the AESA to electronically scan its beam(s) without a mechanical gimbal. They include phase shifters for beam steering, HPAs for transmitting, LNAs for receiving, circulators, filters, and switches. It will be shown in Chapter 4 that TRMs are responsible for ensuring that AESAs have graceful degradation and operate at requirement levels with failures. This is extremely important in terms of system reliability.

The TRMs effect on the SNR equation is on P_{TX} and F. The TRMs are the primary contributor toward maximum radiated power for maximal ERP and also drive the system sensitivity. For large arrays (>100 elements), the TRMs are typically the price driver for an AESA system. Great care is taken to ensure that they are optimally designed for minimized cost and complexity.

1.5.3 Beamformer

The beamformer for an AESA is a passive circuit whose role is to distribute the signal from the exciter to every array element and also to combine the signals received at every array element into a coherent sum beam. Passive beamformers directly affect P_{TX} and F and have to be optimized to minimize loss. Beamformers, as will be shown in Chapter 5, are also important for systems that do geolocation (radar, ESM, SIGINT). Multiple beamformers or a single beamformer with multiple outputs provide the receiver with sum and delta beams to enable geolocation.

1.6 AESA Cascaded Performance and Architecture Selection

Signal and noise gain, cumulative noise figure, cumulative intercept point, and spurious free dynamic range (SFDR) are all key peformance parameters for an AESA. Because of this, it is critical to understand how to calculate these parameters to effectively summarize AESA performance. Great care is taken in the design of an AESA to understand how all of the cascaded electronics operate together for overall AESA performance. Chapter 6 describes how to calculate these parameters with summary expressions to describe them.

Another aspect of AESA design is selecting the right architecture for different applications. As an example, AESAs that have a wide operational frequency bandwidth must be designed with a different architecture than an AESA that has a narrow band requirement. Chapter 7 covers the various AESA architectures that include analog beamforming, SA beamforming, overlapped SA beamforming, SA digital beamforming (DBF), and elemental digital beamforming (EDBF). These architectures provide a menu to select from to satisfy system level requirements such as operational bandwidth, IBW, maximum scan angle, and number of beams. Finally, adaptive beamforming will be discussed, which is a capability that is enabled by the SA DBF and elemental DBF architectures.

References

Balanis, C. *Antenna Theory Analysis and Design*. John Wiley & Sons, Publishers, Inc., 1982.

Grumman, Northrop "Northrop Grumman Demonstrates Antenna Sharing and Pattern Capabilities at Naval Research Laboratory Test Facility." https://news.northropgrumman.com/news/releases/northrop-grumman-demonstrates-antenna-sharing-and-pattern-capabilities-at-naval-research-laboratory-test-facility, 2019.

Bell Labs. "ABM Research and Development at Bell Laboratories, Project History." *Technical Report*, 1975.

Pettai, R. *Noise in Receiving Systems*. John Wiley & Sons, 1984.

Reim, G. "US Navy's Mid-Band Jammer Pod Makes First Flight on Boeing ea-18g Growler." https://www.flightglobal.com/fixed-wing/us-navys-mid-band-jammer-pod-makes-first-flight-on-boeing-ea-18g-growler/139703.article, 2021.

Skolnik, M.I. *Radar Handbook*. McGraw Hill, 1990.

Tokoro, S., Kuroda, K., Kawakubo, A., Fujita, K., and Funinami, H. "Molecular fMRI." *IEEE IV2003 Intelligent Vehicles Symposium Proceedings*, 2003.

2

AESA Theory

Key Concepts

- AESA 1D and 2D Pattern Expressions for Electronic Scanning
- Phase vs. Time Delay Steering
- Beamwidth
- Instantaneous Bandwidth (IBW)
- Grating Lobes
- Error/Quantization Effects
- Circular and Conformal Pattern Synthesis
- Rectangular and Triangular Array Grids
- Coordinate Systems
- Roll, Pitch, and Yaw
- Integrated Gain

2.1 Introduction

AESAs provide the capability of commandable, agile, high gain beams, which is advantageous for applications such as radar, weather surveillance, and imaging. In contrast to reflector antennas, which require a gimbal for steering the array beam, AESAs electronically scan the array beam in space without physical movement of the array (see Figure 2.1). Scanning the beam with an AESA can be performed on the order of nanoseconds as opposed to milliseconds for a reflector. This enables increased scanning rates in addition to flexibility to command the array beam's position in space for tailored modes of operation.

Active Electronically Scanned Arrays: Fundamentals and Applications, First Edition. Arik D. Brown.

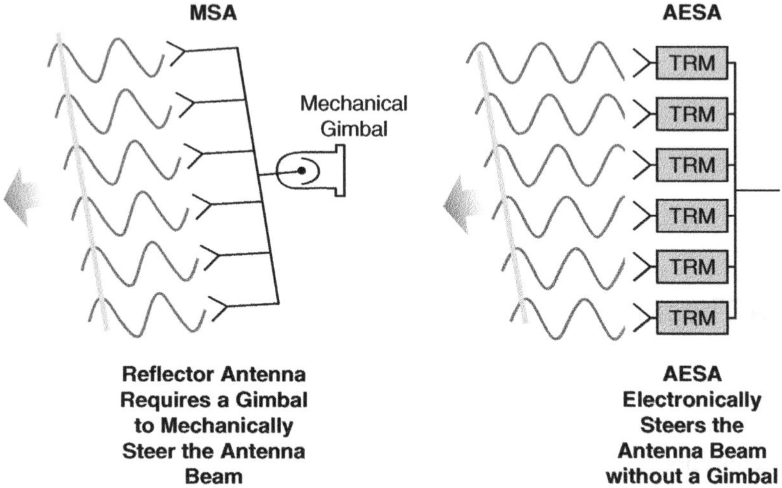

Figure 2.1 Reflector vs. ESA steering.

When designing AESAs, there are basic fundamentals that need to be understood for a successful design (i.e., grating lobes, beamwidth, instantaneous bandwidth, etc.). These fundamental principles will be covered in detail in the remainder of this chapter.

2.2 General One-Dimensional Formulation

2.2.1 Pattern Expression without Electronic Scanning

Consider a one-dimensional array of M elements as shown in Figure 2.2. The elements are uniformly spaced with a spacing of d. The overall length of the array, L, is equal to $M \cdot d$. The elements are centered about $x = 0$, and their position can be denoted as:

$$x_m = (m - 0.5(M + 1))d, \text{where } m = 1, \dots, M. \tag{2.1}$$

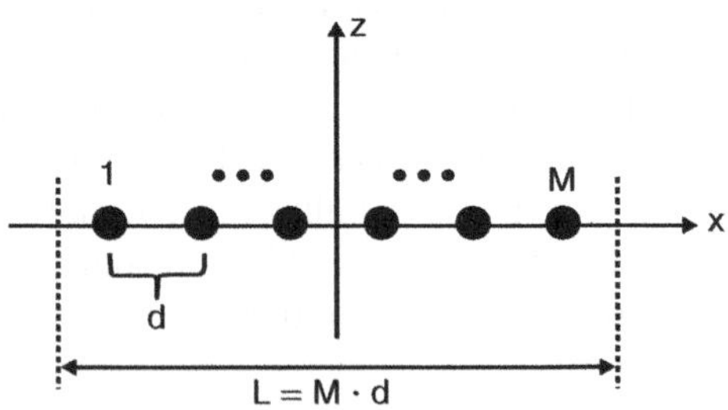

Figure 2.2 Linear array of M elements, with physical spacing *d*. *L* is the length of the array.

Each element has a complex voltage denoted as A_m. A signal that is incident on the array from a direction θ is captured by each of the array elements and is then summed together coherently for the composite signal. The expression for the coherent sum of voltages is represented as:

$$AF = \sum_{m=1}^{M} A_m e^{j\frac{2\pi}{\lambda}x_m sin\theta}. \tag{2.2}$$

AF is the array factor and describes the spatial response of the *M* array elements. From Equation 2.2, we can see that the *AF* is a function of the aperture distribution (A_m), frequency ($\lambda = \frac{c}{f}$, c is the speed of light and f is frequency), element spacing (d), and angle (θ). The *AF* in Equation 2.2 has a maximum value when $\theta = 0°$. This maximum value is M, which is the number of elements in the array. As we'll show later, regardless of whether the array is one- or two-dimensional, the maximum value of the *AF* is always equal to the number of elements in the array.

The *AF* does not completely describe the spatial response of the array. Each of the elements in the array has an element pattern, which is the individual array element's power spatial response. A good expression for modeling the element pattern is a cosine function raised to a power, which is called the element factor (*EF*). The expression for the element pattern (*EP*) is:

$$EP = cos^{\frac{EF}{2}}\theta \tag{2.3}$$

It is important to point out that Equation 2.3 is the voltage expression for the element pattern. The power expression is simply $|EP|^2 = cos^{EF}\theta$. This is the loss that is typically used in the RRE to account for gain loss as a function of scan. In real applications, the *EP* is not zero at $\theta = 90°$, and antenna measurements must be performed in place of the cosine expression in Equation 2.3 for angles near 90°. An AESA, in its installed environment or in a measurement range, will be subject to diffraction and reflections near the edges of the array that will modify the *EP* near the edges.

The complete pattern representation for the array is found using pattern multiplication (Balanis, 1982). Pattern multiplication states that the complete pattern can be calculated by multiplying the *AF* and the *EP*. This assumes that the *EP* is identical for each element in the AESA, which for large AESAs is a good approximation. Equation 2.4 shows the total pattern for the array of *M* elements.

$$F(\theta) = EP \cdot AF = cos^{\frac{EF}{2}}\theta \cdot \sum_{m=1}^{M} A_m e^{j\frac{2\pi}{\lambda}x_m sin\theta}. \tag{2.4}$$

Several important points can be gleaned from Equation 2.4. The first is that the expression for the *EP* has been factored out of the *AF* under the assumption that the element pattern is the same for all elements. As we will see later in the chapter, this is not true for conformal arrays where the normal of the element pattern is not parallel with the normal of the array boresite ($\theta = 0°$) and is also incorrect for arrays with a small number of elements. However, outside of the discussion on conformal arrays, for the remainder of this chapter all AESAs described will assume a large number of elements ($> 5\lambda$) (Holter and Steyskal, 2002).

2.2.2 Pattern Expression with Electronic Scanning

In the previous section a general result was shown for the spatial pattern representation of a linear M-element array. In this section we will express the pattern with the inclusion of electronic scan. The expression in Equation 2.4 only has a maximum value when $\theta = 0°$. An AESA has the ability to scan the beam so that the beam has a maximum at other angles ($\theta \neq 0°$). For the remainder of this book the scan angle will be denoted as θ_o.

Scanning the beam of the array requires adjusting the phase and/or time delay of each element in the array. By rewriting the expression in Equation 2.2 and expanding the complex voltage at each element ($A_m = a_m e^{j\theta_m}$) Equation 2.2 becomes:

$$AF = \sum_{m=1}^{M} a_m e^{j\theta_m} e^{j\frac{2\pi}{\lambda}x_m sin\theta}. \tag{2.5}$$

The AF then has a maximum at θ_o when $\theta_m = -\frac{2\pi}{\lambda}x_m sin\theta_o$. Equation 2.5 can then be expressed as:

$$AF = \sum_{m=1}^{M} a_m e^{j\left(\frac{2\pi}{\lambda}x_m sin\theta - \frac{2\pi}{\lambda}x_m sin\theta_o\right)}. \tag{2.6}$$

By applying the appropriate phase at each element, the AESA beam can be electronically scanned spatially without physically moving the entire array. This is the excitement of AESAs! The overall pattern can now be expressed as:

$$F(\theta) = cos^{\frac{EF}{2}}\theta \cdot \sum_{m=1}^{M} a_m e^{j\left(\frac{2\pi}{\lambda}x_m sin\theta - \frac{2\pi}{\lambda}x_m sin\theta_o\right)}. \tag{2.7}$$

Electronic scan can be categorized as phase steering or time delay steering. For phase steering, each element has a phase shifter and applies the appropriate phase as a function of frequency and scan angle. A characteristic of phase shifters is that their phase delay is designed to be constant over frequency. This means the expression in Equation 2.7 must be modified to account for this. The pattern expression for phase delay steering becomes:

$$F(\theta) = cos^{\frac{EF}{2}}\theta \cdot \sum_{m=1}^{M} a_m e^{j\left(\frac{2\pi}{\lambda}x_m sin\theta - \frac{2\pi}{\lambda_o}x_m sin\theta_o\right)}, \tag{2.8}$$

where $\lambda = \frac{c}{f}$ and $\lambda_o = \frac{c}{f_o}$. What is readily seen is that when $f \neq f_o$ the pattern is no longer a maximum. This will be discussed in more depth in Section 2.3.2.

When time delay is used, Equation 2.8 becomes:

$$F(\theta) = cos^{\frac{EF}{2}}\theta \cdot \sum_{m=1}^{M} a_m e^{j\frac{2\pi}{\lambda}x_m(sin\theta - sin\theta_o)}. \tag{2.9}$$

2.3 AESA Fundamental Topics

The AF for a one-dimensional AESA was shown to be:

$$AF = \sum_{m=1}^{M} a_m e^{j(\frac{2\pi}{\lambda}x_m sin\theta - \frac{2\pi}{\lambda}x_m sin\theta_o)}. \tag{2.10}$$

Mathematically, Equation 2.10 can be represented in a closed-form solution, and the alternate expression for the AF is:

$$AF = \frac{sin[M\pi d(\frac{sin\theta_o}{\lambda_o} - \frac{sin\theta}{\lambda})]}{sin[\pi d(\frac{sin\theta_o}{\lambda_o} - \frac{sin\theta}{\lambda})]}. \tag{2.11}$$

Equation 2.11 is the phase shifter representation of the AF. The corresponding time delay formulation is:

$$AF = \frac{sin[\frac{M\pi d}{\lambda}(sin\theta_o - sin\theta]}{sin[\frac{\pi d}{\lambda}(sin\theta_o - sin\theta]}. \tag{2.12}$$

The derivation for the closed-form solution of the AF can be found in Appendix A. The next three topics covered are readily derivable from Equations 2.11 and 2.12.

2.3.1 Beamwidth

The AESA beamwidth refers to the angular extent of the main beam where the power drops by a certain value. When that value is 3 dB, the beamwidth can also be referred to as the half-power beamwidth. The common expression used for back-of-the-envelope calculations, $\frac{\lambda}{L}$, will be shown to be the 4 dB beamwidth for a uniform distribution. The 4 dB expression is commonly used as it provides a good estimate for calculations not requiring precision.

Equation 2.12 has the form of $\frac{sinMx}{x}$ which can be approximated by $\frac{sinx}{x}$, the standard sinc function. The 4 dB point for the sinc function occurs when $x = \pm\frac{\pi}{2}$. When $x = \pm\frac{\pi}{2}$, $\frac{sinx}{x} = \frac{2}{\pi}$. With this same logic the expression in Equation 2.12 can be written as:

$$AF \approx \frac{sin[M\pi d(\frac{sin\theta_o}{\lambda_o} - \frac{sin\theta}{\lambda})]}{M\pi d(\frac{sin\theta_o}{\lambda_o} - \frac{sin\theta}{\lambda})}. \tag{2.13}$$

Setting the arguments in Equation 2.13 equal to $\pm\frac{\pi}{2}$ when $\theta = \theta_o \pm \frac{\theta_{BW}}{2}$ produces the following two equations:

$$M\pi d\left(\frac{sin\theta_o}{\lambda_o} - \frac{sin(\theta_o + \frac{\theta_{BW}}{2})}{\lambda}\right) = \frac{\pi}{2} \tag{2.14}$$

and

$$M\pi d\left(\frac{sin\theta_o}{\lambda_o} - \frac{sin(\theta_o - \frac{\theta_{BW}}{2})}{\lambda}\right) = -\frac{\pi}{2}. \tag{2.15}$$

Subtracting Equation 2.15 from Equation 2.14 and applying a trigonometric identity produces the expression:

$$\frac{M\pi d}{\lambda}\left(2sin\frac{\theta_{BW}}{2}cos\theta_o\right) = \pi. \tag{2.16}$$

Using the sine small-angle approximation Equation 2.16 can be solved for the beamwidth and is expressed as

$$\theta_{BW} = \frac{\lambda}{Mdcos\theta_o} = \frac{\lambda}{Lcos\theta_o}, \tag{2.17}$$

where $L = M \cdot d$, and is the length of the AESA aperture. Equation 2.17 is valid for both phase shifter and time delay steering, and for $\theta_o = 0$ it is the typical equation used to estimate AESA beamwidth. From Equation 2.17 we can see that the beamwidth is inversely proportional to frequency, aperture length, and the cosine of the scan angle. The beamwidth in Equation 2.17 is the 4 dB beamwidth for a uniform aperture illumination. A more general expression for the beamwidth is

$$\theta_{BW} = \frac{k\lambda}{Lcos\theta_o}, \tag{2.18}$$

where k is the beamwidth factor and varies depending on the aperture distribution. As an example, $k = 0.886$ for the 3 dB beamwidth of a uniformly illuminated AESA. Figure 2.3 shows a plot of AESA beamwidth as a function of scan angle and frequency for $k = 1$ and constant aperture length L.

2.3.2 Instantaneous Bandwidth

When describing instantaneous bandwidth (IBW), it is advantageous to first think of it from a phase shifter perspective. For an AESA employing phase delay, the phase shifters are set at each element to scan the beam. The phase shifters have the characteristic of constant phase vs. frequency. At the tune frequency of the AESA, $f = f_o$ in Equation 2.11, and the AF has a maximum value. When $f = f + \Delta f_o$, the AF no longer has a maximum value at $f = f_o$, and there is an associated pattern

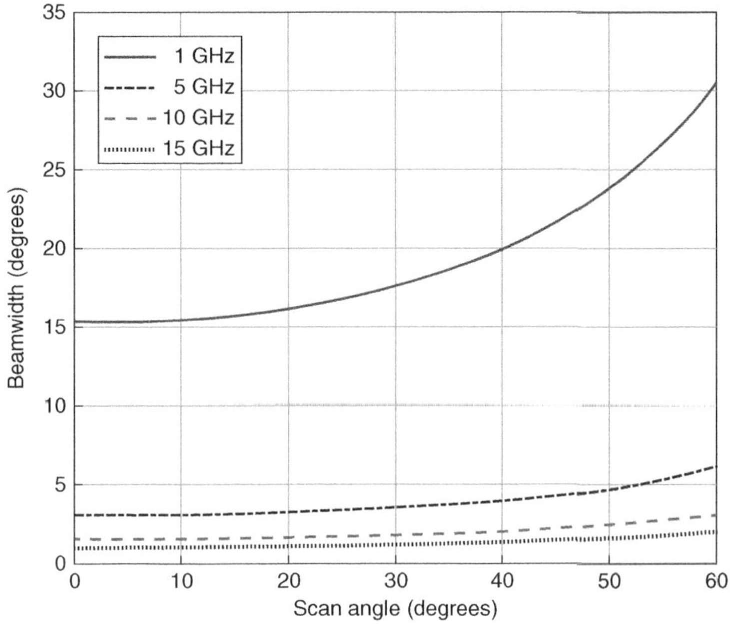

Figure 2.3 AESA beamwidth as a function of scan angle and frequency for $k = 1$.

loss at the commanded scan angle. This phenomenon is commonly referred to as beam squint. The IBW is the range of frequencies over which the loss is acceptable and is $2\Delta f$. Typically, the IBW specified is the 3 dB or 4 dB IBW. Figure 2.4 illustrates beam squint for two AESAs with different aperture lengths. From the plots in Figure 2.4 we can see that the array with the longer length (bottom plot in Figure 2.4) has a smaller beamwidth and suffers more loss at the commanded scan angle of 30°.

The IBW can be derived from Equation 2.13 in a similar fashion to the array beamwidth. (An alternate derivation using the exponential form of the *AF* can be found in Appendix B.) The first step is to express Equation 2.13 in terms of frequency:

$$AF \approx \frac{sin[\frac{M\pi d}{c}(f_o sin\theta_o - f sin\theta)]}{\frac{M\pi d}{c}(f_o sin\theta_o - f sin\theta)}. \tag{2.19}$$

Substituting $f = \Delta f_o$ into 2.19 and solving for Δf, the following expression is obtained (See Appendix B):

$$IBW = \Delta f = \frac{c}{Mdsin\theta_o} = \frac{c}{Lsin\theta_o}. \tag{2.20}$$

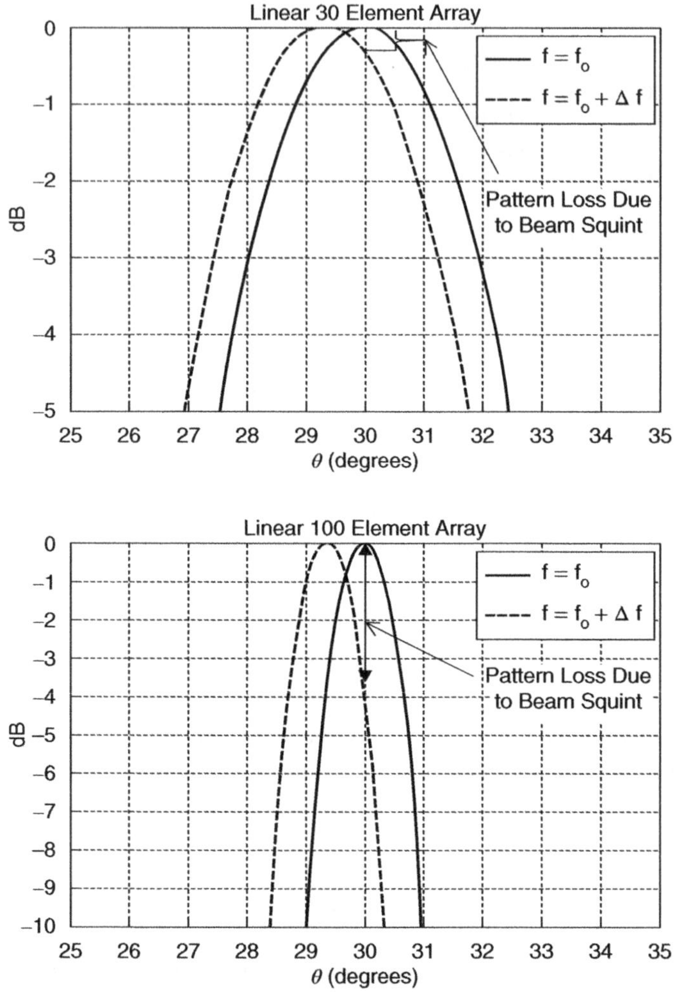

Figure 2.4 Beam squint using phase shifter steering for AESAs with the same element spacing with differing number of elements. In both plots the solid line pattern represents operation at the tune frequency f_o and the dashed line represents beam squint at $f = f_o + \Delta f$.

Equation 2.20 defines the 4 dB IBW for an AESA. Similar to Equation 2.18, the IBW can also be written more generally as

$$IBW = \frac{kc}{Lsin\theta_o}, \tag{2.21}$$

where k is the beamwidth factor, which is a function of the aperture distribution. For the case of time delay steering, there is no beam squint. Looking at

Equation 2.12, we can see that for time delay, the AF has a maximum value at the commanded scan angle. This makes time delay very attractive for wide bandwidth applications and also for large arrays that have a small beamwidth and have limited bandwidth using phase delay steering.

2.3.3 Grating Lobes

The AF is a periodic function. Similar to signal processing theory, the elements in a phased array if not sampled properly with the correct element spacing will generate grating lobes, which are simply periodic copies of the main beam. The grating lobe locations are a function of frequency and element spacing. Grating lobes are problematic because they rob energy from the main beam, which reduces sensitivity and indirectly angle accuracy. Equation 2.11 can be used to derive the grating lobe locations for an AESA. The AF has maximum values when $\pi d(\frac{sin\theta_o}{\lambda_o} - \frac{sin\theta}{\lambda}) = \pm P\pi$ where $P = 0, 1, 2, \ldots$ Rearranging terms, this can be represented as:

$$sin\theta_{GL} = \frac{\lambda}{\lambda_o} sin\theta_o \mp P\frac{\lambda}{d}. \tag{2.22}$$

The first term on the right-hand side of Equation 2.22 represents the location of the main beam when scanned to θ_o. The second term represents the grating lobe locations. Simplifying Equation 2.22 by setting $\lambda = \lambda_o$ and setting $\theta_o = 0°$, we find the expression:

$$sin\theta_{GL} = \mp P\frac{\lambda}{d}, \tag{2.23}$$

which gives the grating lobe locations when the main beam is not scanned (at boresight). To find the element spacing required for grating lobe free scanning to 90° we set $\theta_{GL} = 90°$ and set $P = 1$ in Equation 2.22. This results in the following equation:

$$d = \frac{\lambda}{1 + sin\theta_o}. \tag{2.24}$$

This shows that to scan the AESA to $\theta_o = 90°$, the elements spacing must be $\frac{\lambda}{2}$. Figure 2.5 shows a plot of an AESA pattern, AF, and element pattern with element spacing of λ.

The grating lobes of the AF appear at 90°, which is undesirable. However, because of pattern multiplication, the roll-off of the element pattern attenuates the grating lobe of the AF. Figure 2.6 shows pattern plots overlaid with element spacing of $\frac{1}{2}\lambda, 1\lambda$, and 2λ. A well-designed AESA takes into account grating lobes ensuring that they are not present within the required electronic scan volume. This will be discussed in the context of sine space in Section 2.6 for two-dimensional AESAs.

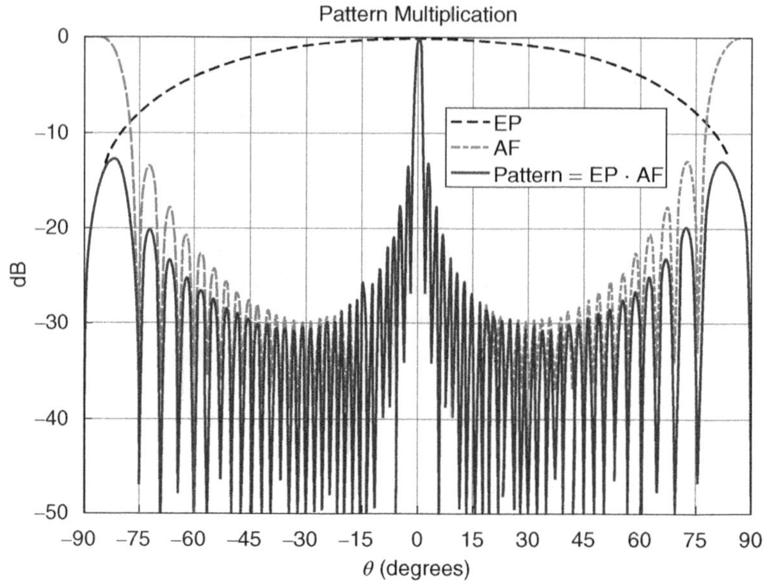

Figure 2.5 The overall pattern can be computed using pattern multiplication: The element pattern of a single representative element in an array environment multiplied by the array factor.

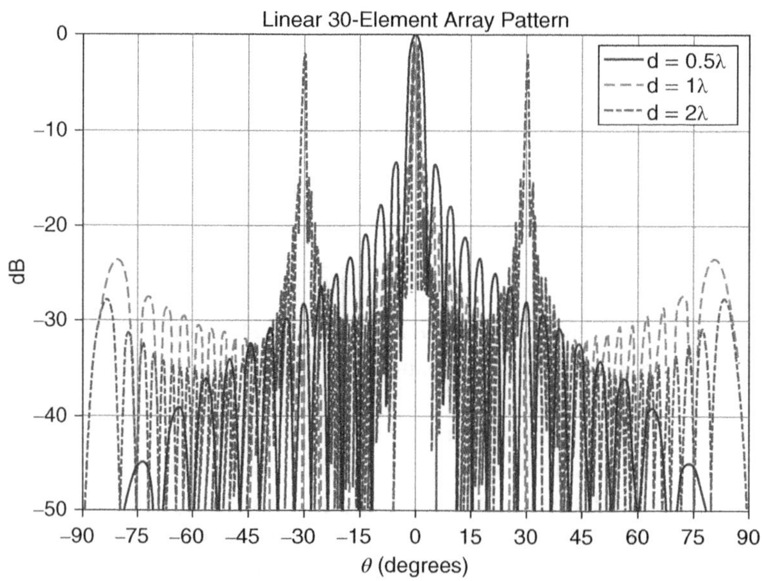

Figure 2.6 Grating lobes are a function of the element spacing in an AESA. The element spacing is optimized in the AESA design to prevent grating lobes, which degrade performance.

2.3.4 Error Effects

Many errors are introduced in an actual AESA design. These errors are both random and correlated and are caused by imperfect components and signal distribution networks (Mailloux, 1993). If these errors are not controlled, they can result in large sidelobe levels, which is undesired. The primary source of correlated errors is due to quantization of the phase shifters (or time delay units) and attenuators used to implement electronic scan and amplitude taper in an AESA. An additional source of correlated error is time delay implemented at the subarray level for large wideband AESAs (Mailloux, 1993). Random amplitude and phase errors in an AESA can be attributed to things such as failed elements and manufacturing tolerances. In order to achieve good sidelobe level performance, both quantization and random errors should be simulated.

2.3.5 Quantization Effects

In an AESA, phase shifters (and/or time delay units) provide the progressive phase across the array required to scan the beam. A phase shifter is usually characterized by its number of bits (N). The least significant bit (LSB) is then calculated as

$$LSB = \frac{360^\circ}{2^N}. \tag{2.25}$$

By quantizing the phase at each element, the phase across the array is implemented as a staircase approximation to the required phase. This is illustrated in Figure 2.7. This produces a periodic triangular phase error, which leads to undesired sidelobes with a grating lobe–like periodicity (Mailloux, 1993). Miller (Miller, 1964) derived expressions for the peak and average SLL due to quantization error only, which are

$$\begin{aligned} AverageSLL &= \frac{1}{3n_{elem}\epsilon}\frac{\pi^2}{2^{2N}}, \\ PeakSLL &= \frac{1}{2^{2N}}. \end{aligned} \tag{2.26}$$

The quantization of amplitude taper has a similar effect as mentioned with phase. Instead of a smooth aperture amplitude distribution, the amplitude is quantized by N bit attenuators. This appears as another source of error. Figure 2.8 shows the effects of quantization. The pattern with 2 bit phase shifters and attenuators has significant sidelobes due to quantization error as compared to the ideal pattern. Figure 2.9 shows the comparison of an ideal pattern with that of the pattern employing 6 bit phase shifters and attenuators. The pattern with 6 bit element control has significant performance improvement over the pattern with only 2 bits. The quantization error raises the average SLL but not enough for significant SLL

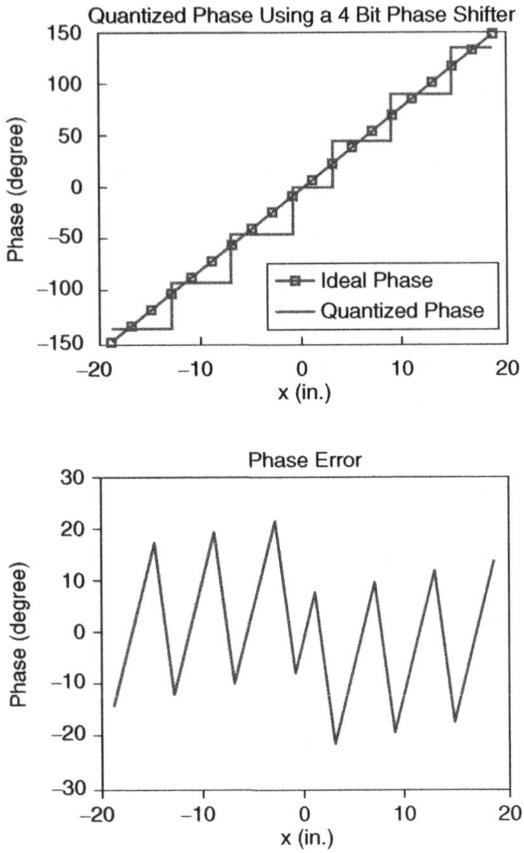

Figure 2.7 Staircase approximation for a 4 bit phase shifter.

effect employing 6 bit phase shifters and attenuators. In most AESA applications, using 6 bits for phase and 6 bits for attenuation is sufficient for satisfying the required performance. Additional discrete bits for attenuation are sometimes used for active gain control in the RF chain.

2.3.6 Random Error Effects (Amplitude and Phase)

Random errors that occur, such as failed elements, can be modeled as having a Gaussian distribution with variance and zero mean for both phase and amplitude (Mailloux, 1993). The average SLL due to random errors can be expressed as (Mailloux, 1993)

$$\overline{\sigma^2} = \frac{(\pi)^{\frac{1}{2}}\overline{\epsilon^2}}{D_A^{\frac{1}{2}}P}, \tag{2.27}$$

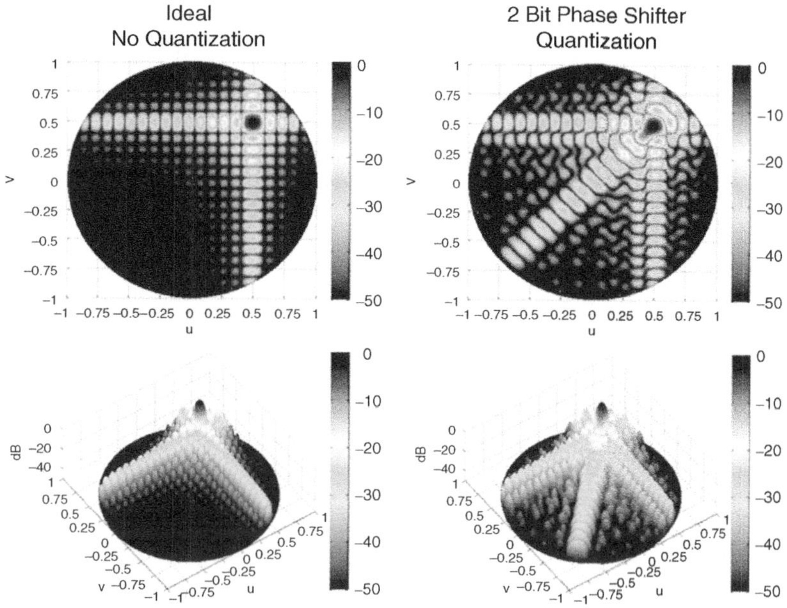

Figure 2.8 Comparison of an ideal pattern and a pattern with 2 bit phase shifters and attenuators. Note the quantization sidelobes that appear due to using insufficient bit resolution.

where $\overline{\sigma^2}$ is the average SLL, D_A is the directive gain, P is the probability of an element working (not failed), and $\overline{\epsilon^2}$ is the error variance, which is a function of the phase and amplitude errors, which have a Gaussian distribution. Figure 2.10 shows a plot of 6° phase and 0.5 dB amplitude random errors (1σ). The phase and amplitude error values are typically specified to a performance requirement. Figure 2.11 illustrates the effects of the random errors on the pattern as compared to that of an ideal pattern.

2.4 One-Dimensional Pattern Synthesis

Now that we have derived the pattern expression for a one-dimensional AESA, we will look at pattern plots while varying the element amplitude distribution, frequency, number of elements, and scan angle. We'll begin by plotting the *EP*, *AF*, and pattern all on a single plot to visualize pattern multiplication. (See Figure 2.12.)

The *EF* in this example is 1.5. The *EF* is dependent on the radiator design; however, 1.5 is a safe number to use for analysis. In system design, the power pattern

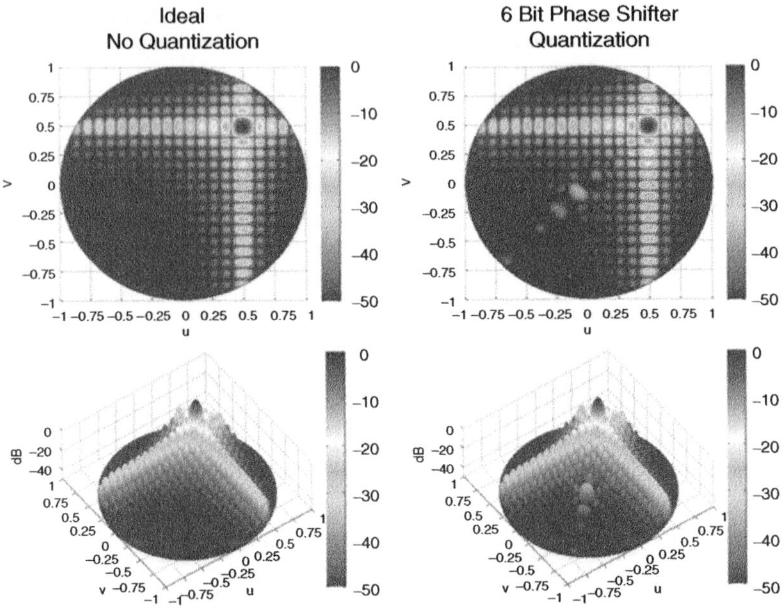

Figure 2.9 Comparison of an ideal pattern and a pattern with 6 bit phase shifters and attenuators. The performance in this example utilizing 6 bits approximates the ideal pattern well enough to use for actual implementation. Adding more bits would unnecessarily increase the cost and complexity of the AESA.

is what determines performance, and all figures will be plotted as power patterns. In Figure 2.12 all the power patterns are plotted in decibels (dB). Equation 2.28 shows how the power pattern is computed in dB.

$$F_{dB} = 10log_{10}(EP \cdot AF)^2 = 20log_{10}(EP) + 20log_{10}(AF). \tag{2.28}$$

Figure 2.13 shows the overlay of the *EP*, *AF*, and array patterns with the array scanned. The figure illustrates that the *EP* does not scan with the *AF*. The *EP* roll-off attenuates the *AF* pattern. This roll-off is governed by the *EF*. As an example, if the *EF* equaled one, an additional loss of 3 dB would be added to the array pattern at $\theta = 60°$ due to the *EP* ($20 \cdot log_{10}(cos60°) \approx -3$ dB). This difference has the most effect at large scan angles where the *EP* scan loss must be accounted for in analyzing performance. In addition to scan loss, the *EP* causes the array pattern peak to be shifted from the scan angle. Figure 2.14 shows a zoomed-in view of the *EP*, *AF*, and array pattern. In the figure it is shown that the *AF* has its peak value at the scan angle of 60°; however, the array pattern is slightly shifted because of the *EP* roll-off.

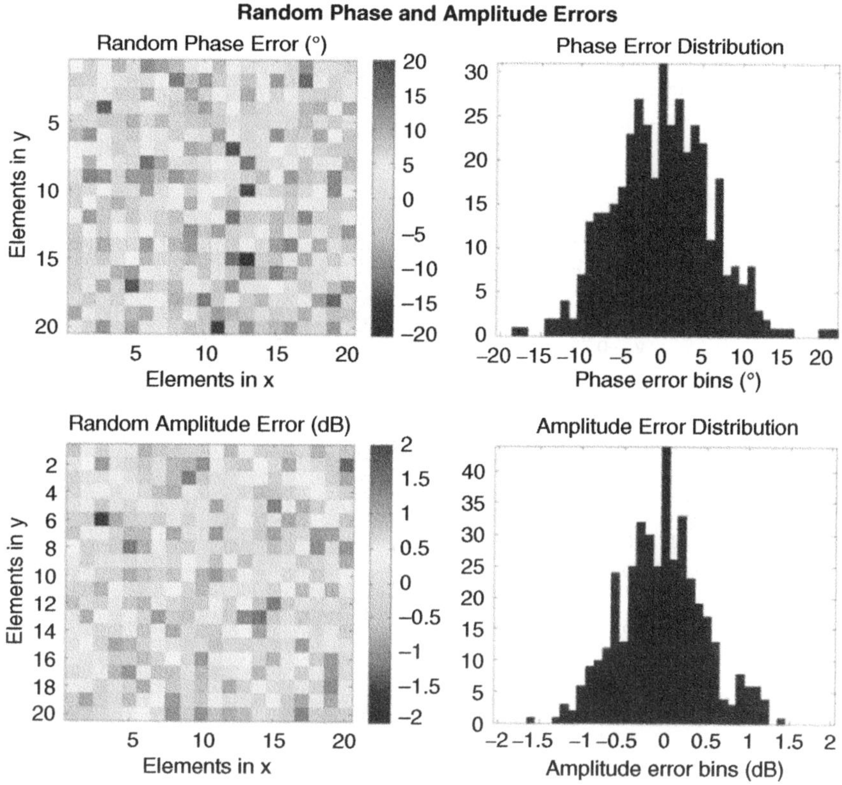

Figure 2.10 Plot of Gaussian distributed random phase and amplitude errors with a standard deviation of 6° and 0.5 dB, respectively.

2.4.1 Varying Amplitude Distribution

In Section 2.2.2, it was shown that by varying the phase of each element in the array the AESA beam can be scanned spatially. In addition to modifying the element phase, the element amplitudes (A_m) can be modified as well to lower sidelobe levels. Figure 2.15 shows the pattern for a uniform amplitude distribution across the array ($A_m = 1$). For a uniform distribution the first sidelobes are 13 dB below the peak of the main beam. When the AESA beam is scanned, the sidelobe level, relative to the main beam, changes. This is due to the roll-off of the *EP* affecting the array pattern. In many AESA applications it is undesirable for the main beam and the sidelobes to be close in magnitude. Signals incident on the AESA with a high enough power level could appear as signals that came in the main beam. Figure 2.16 shows an array scanned to 60°. In this example, the first sidelobe is now only 10 dB lower than the main beam.

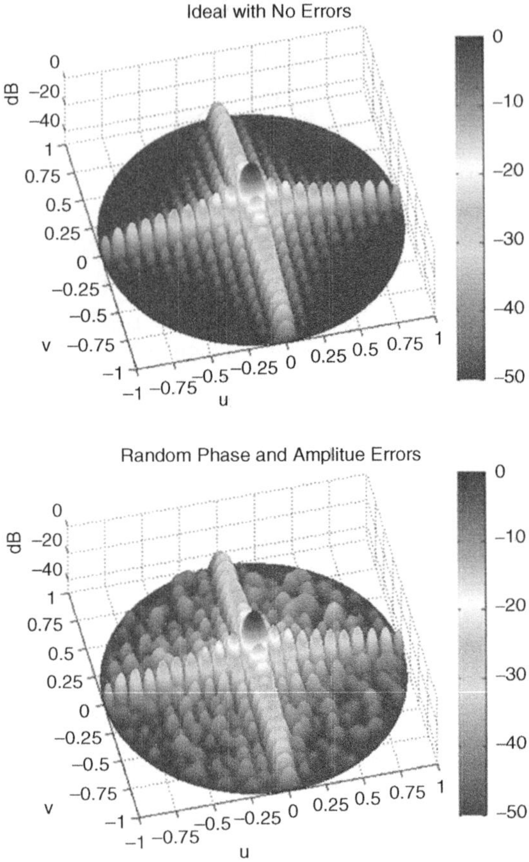

Figure 2.11 Comparison of an ideal pattern with a pattern that has the random phase and amplitude errors shown in Figure 2.10.

In order to reduce the sidelobe level of the AESA, amplitude weighting can be applied. Various types of weightings can be used similar to filter theory; however, Taylor weighting is the most efficient aperture distribution (Taylor, 1955). Figure 2.17 illustrates the difference between the uniform illumination and the Taylor weighting illumination. There is a loss associated with weightings that deviate from uniform weighting, which is the most efficient. This will be discussed later in the section on beamwidth. The pattern for the array modeled in Figure 2.15 using Taylor weighting is plotted in Figure 2.18. A 30 dB Taylor weighting is applied. The figure shows the sidelobes less than 30 dB below the main beam. In practice, there are amplitude errors across the array, which will cause some of the sidelobes to be above 30 dB. Typically an average sidelobe level is specified. The effects of errors on the AESA pattern will be discussed in Section 2.3.4. Figure 2.19 depicts the AESA with 30 dB Taylor weighting

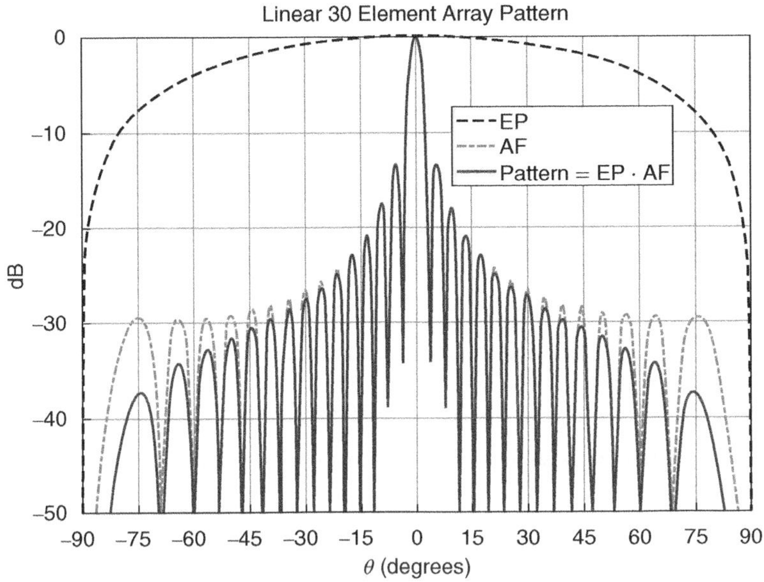

Figure 2.12 EP, AF, and pattern overlaid.

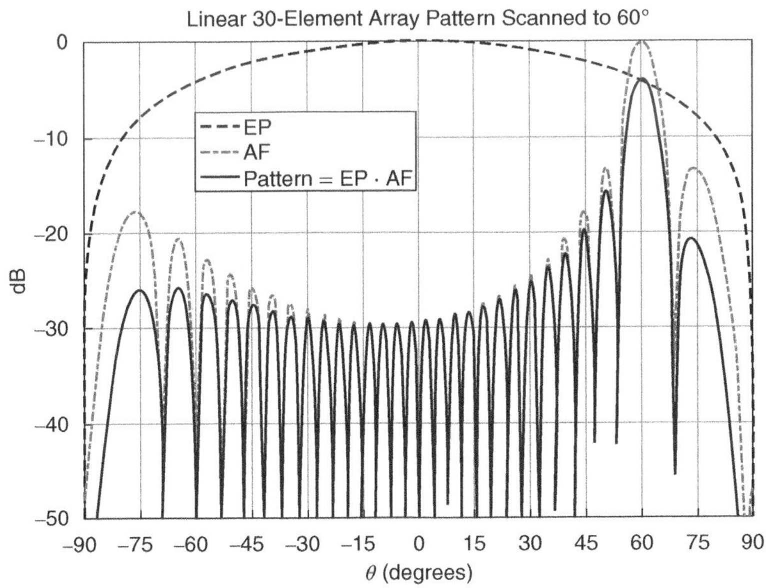

Figure 2.13 EP, AF, and pattern overlaid with scan.

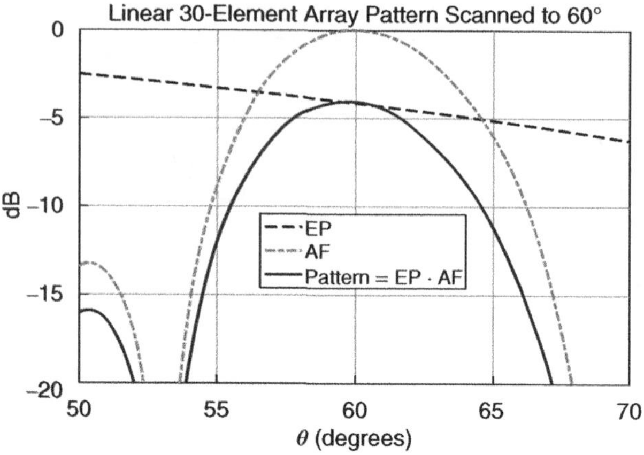

Figure 2.14 Pattern peak shifted from the scan angle due to *EP* roll-off.

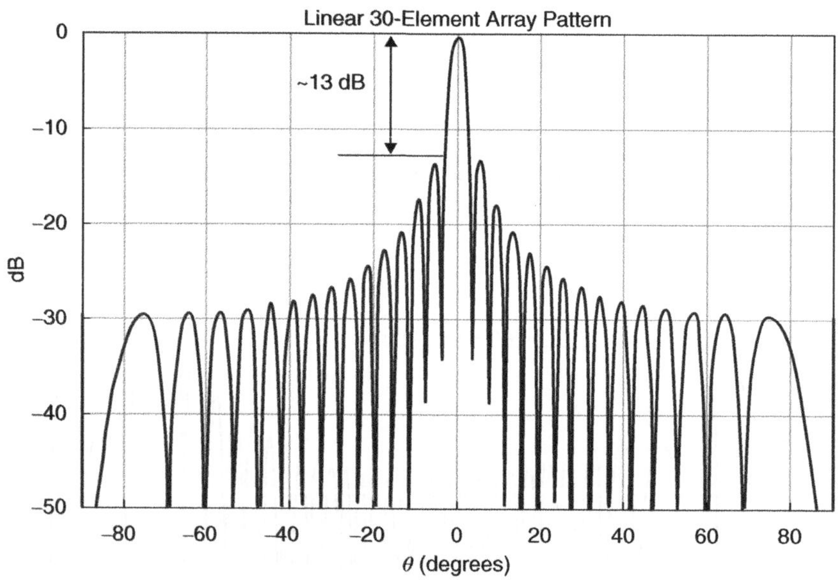

Figure 2.15 Pattern with uniform weighting.

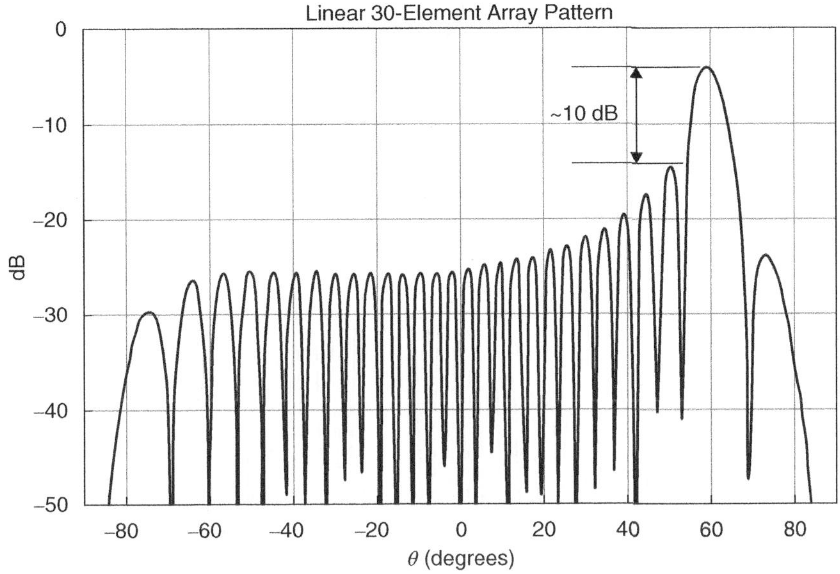

Figure 2.16 Pattern with uniform weighting scanned to 60°.

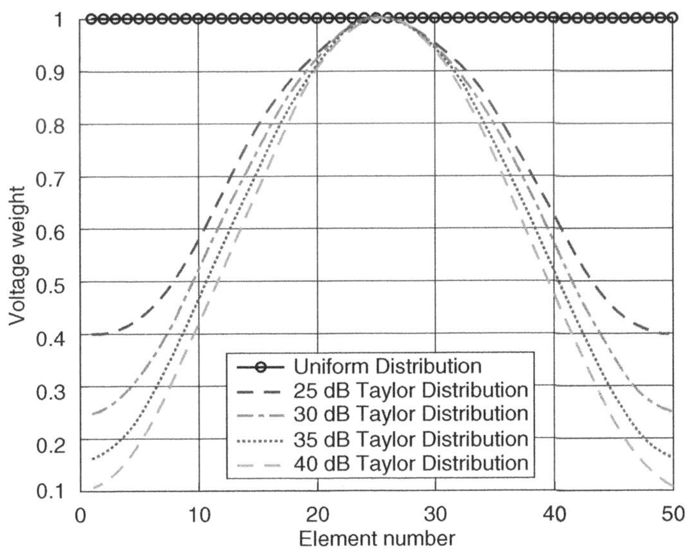

Figure 2.17 Uniform vs. Taylor weighting.

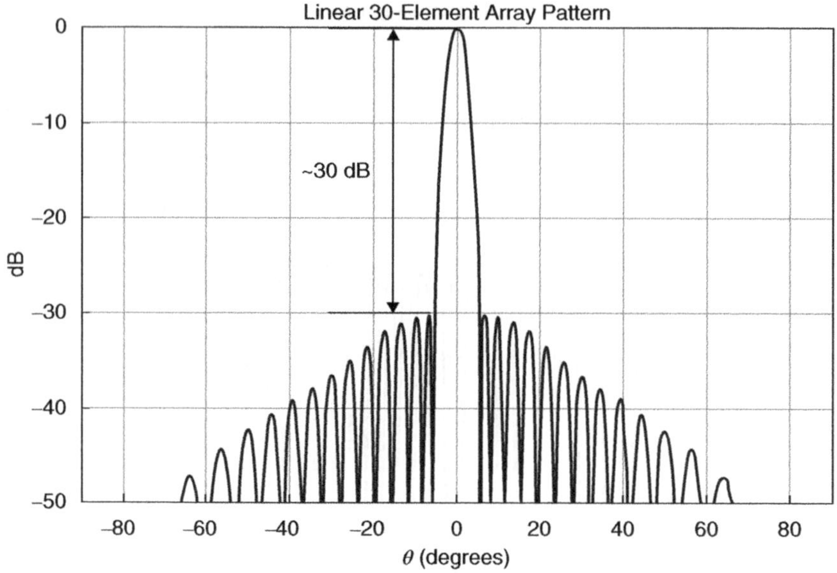

Figure 2.18 Pattern with Taylor weighting.

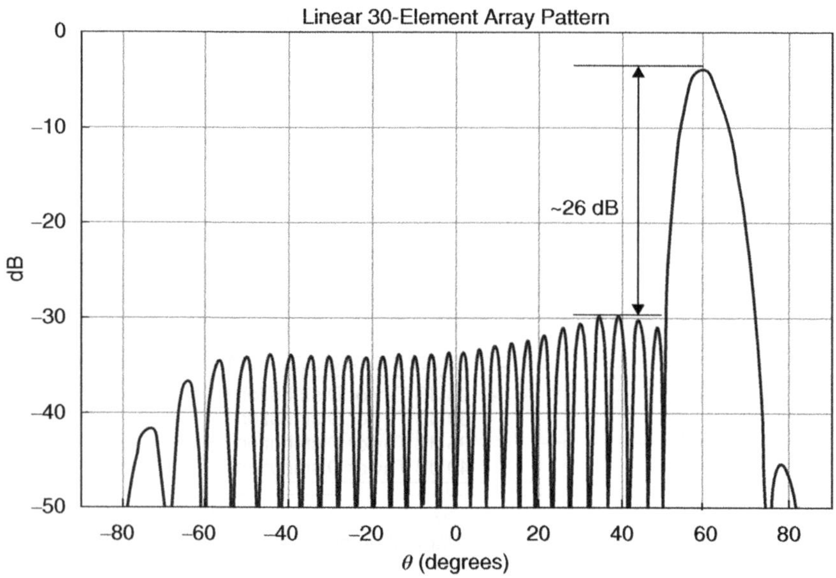

Figure 2.19 Pattern with Taylor weighting scanned to 60°.

scanned to 60°. In contrast to Figure 2.16, the sidelobes here are ≈ 26 dB below the main beam as opposed to 10 dB in the uniform weighting case.

2.4.2 Varying Frequency

As the operating frequency of the AESA is changed, the beamwidth and the sidelobes are changed as well. As discussed previously, the beamwidth is inversely proportional to frequency. In a practical application the AESA will be required to function over an operational frequency range. As the frequency is changed over the operational frequency band, the beamwidth will change in size. At the high end of the band the beamwidth will be the smallest, and at the low end of the band the beamwidth will be the largest. Figure 2.20 shows the patterns of a 1D AESA at three different frequencies to demonstrate the change in beamwidth as a function of frequency. In Figure 2.20, uniform weighting was used in order to highlight the change in beam shape and sidelobe orientation and size.

2.4.3 Varying Scan Angle

One of the great features of an AESA is the ability to electronically scan the beam. However, there is no free lunch. One of the primary effects of electronic scan is the array pattern loss due to the *EP* roll-off which has been previously discussed. However, an additional impact of electronic scan is the broadening of the main beam. As the beam is scanned, the beam broadens at the rate of $\frac{1}{cos\theta}$. This changes the spatial footprint of the main beam and has to be accounted for at a system level.

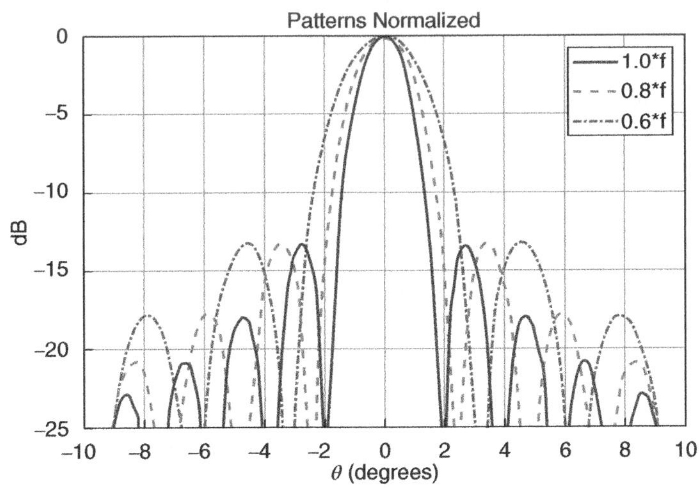

Figure 2.20 Changing beamwidth as a function of frequency.

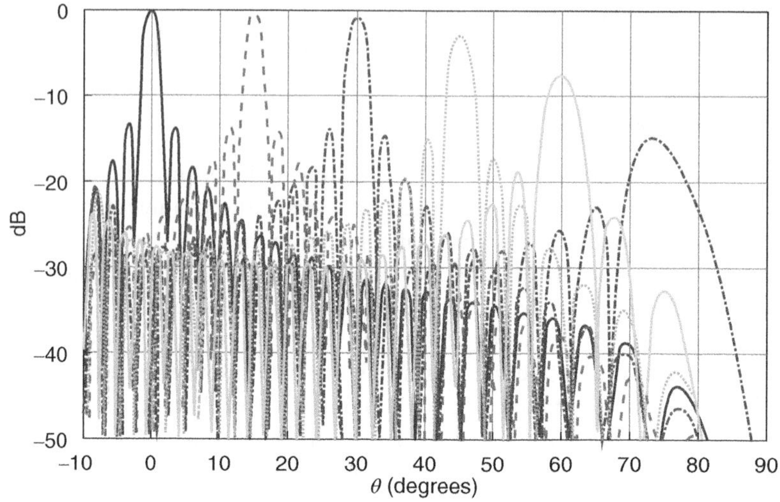

Figure 2.21 Beam broadening with electronic scan.

Figure 2.21 shows an AESA pattern at several different electronic scan angles. The broadening of the main beam with increasing electronic scan is shown.

2.5 Conformal Arrays[1]

For applications involving conformal phased arrays, the normal modeling done for a planar array is no longer valid. In order to characterize the conformal array in terms of pattern distribution, sidelobe levels, and gain performance, it is important to understand how to model a non-planar array. Since the element pattern for each element in a conformal array points in a different direction, each element contributes a different amount of energy to the main beam and sidelobes thereby affecting the pattern.

2.5.1 Array Pattern for a Linear Array

To understand how the pattern equation applies to a conformal array, it is best to start with a more general expression for the pattern:

$$F(\mathbf{r}) = \sum_{i}^{N} a_i EP_i(\theta, \phi) e^{jk\mathbf{r}_i \cdot \hat{\mathbf{r}}}. \tag{2.29}$$

Equation 2.29 is a summation over all of the elements in the array, where N is the number of array elements. Each element has an amplitude, a_i, element pattern

1 This section is based on a technical memo written by the author and Dr. Sumati Rajan with the consultation of Dr. Daniel Boeringer.

EP_i, and phase $k\mathbf{r}_i \cdot \hat{\mathbf{r}}$, where k is the free space wave number, $\mathbf{r}_i$ is the element position vector, and $\hat{\mathbf{r}}_i$ is the spatial unit vector. It is important to note that no phase shift has been applied to the elements (i.e., no steering of the antenna beam). In order to account for beam steering, an appropriate phase term must be added to Equation 2.29. The modified pattern equation is shown below:

$$F(\mathbf{r}) = \sum_{i}^{N} a_i EP_i(\theta, \phi) e^{jk\mathbf{r}_i \cdot \hat{\mathbf{r}}} e^{-jk\mathbf{r}_i \cdot \hat{\mathbf{r}}_o}. \tag{2.30}$$

In Equation 2.30, $\hat{\mathbf{r}}_0$ is the unit scan vector, which corresponds to the angle in space to which the antenna beam is steered. It is important to note that the phase term, in the second exponential in Equation 2.30, represents the phase set by phase shifters in an AESA. When analyzing planar arrays, the element pattern, EP_i, can be moved outside of the summation in Equation 2.30. Figure 2.22 depicts a planar linear array with N elements to illustrate the separability of the element factor for a planar array. The unit normal for each element in Figure 2.22 points in the same direction, and the element pattern for each element is the same. This allows Equation 2.30 to be modified as:

$$F(\mathbf{r}) = EP(\theta, \phi) \sum_{i}^{N} a_i e^{jk\mathbf{r}_i \cdot \hat{\mathbf{r}}} e^{-jk\mathbf{r}_i \cdot \hat{\mathbf{r}}_o}. \tag{2.31}$$

Equation 2.31 can be recognized as the well-known pattern multiplication equation for an array, in which the pattern is equal to the multiplication of the element pattern and the array factor. In most array applications the element pattern, EP, is assumed to be a cosine function raised to a power called the element factor (EF). This is shown in Equation 2.32:

$$EP(\theta, \phi) = \cos^{\frac{EF}{2}}(\theta). \tag{2.32}$$

Equation 2.32 assumes that the elements in the array are located in the xy-plane and that the z direction is normal to the face of the array.

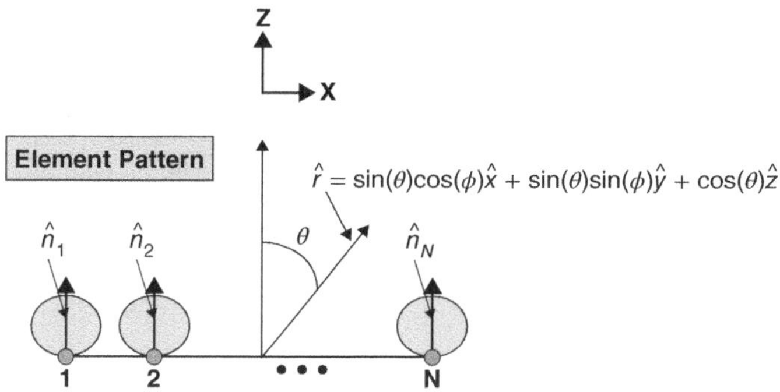

Figure 2.22 Unit normal vectors for the element patterns in a linear array.

To better understand how the element pattern is applied in the case of a conformal array, it is helpful to express the element pattern in a different form,

$$EP(\theta,\phi) = (\hat{n}\cdot\hat{r})^{\frac{EF}{2}}. \tag{2.33}$$

Equation 2.33 makes no assumption on the direction in which the array is pointed. However, when $\hat{n}=\hat{z}$, and using the definition for $\hat{r}$ ($\hat{r} = sin(\theta)cos(\phi)\hat{x} + sin(\theta)sin(\phi)\hat{y} + cos(\theta)\hat{z}$), Equation 2.33 reduces to Equation 2.32. The normal vector in Equation 2.33 is normal to the surface of the array for each element.

2.5.2 Array Pattern for a Conformal Array

When modeling an array of conformal elements, Equation 2.31 is no longer valid. This is because the unit normal for each array element is oriented in a different direction, and the element pattern cannot be removed from the summation in Equation 2.30. This is depicted in Figure 2.23. For computation purposes, the normal for each element must be determined in order to properly compute the antenna pattern. Once the unit normal is known, it can be plugged into Equation 2.33 to calculate the element pattern for each individual element. Substituting Equation 2.33 into Equation 2.30 gives the following expression for the antenna pattern,

$$F(\mathbf{r}) = \sum_{i}^{N} a_i(\hat{n}\cdot\hat{r})^{\frac{EF}{2}} e^{jk\mathbf{r}_i\cdot\hat{\mathbf{r}}} e^{-jk\mathbf{r}_i\cdot\hat{\mathbf{r}}_o}. \tag{2.34}$$

Equation 2.34 shows that for every angle in space (θ, ϕ), each element contributes a different amount of power to the antenna pattern. This concept is illustrated in

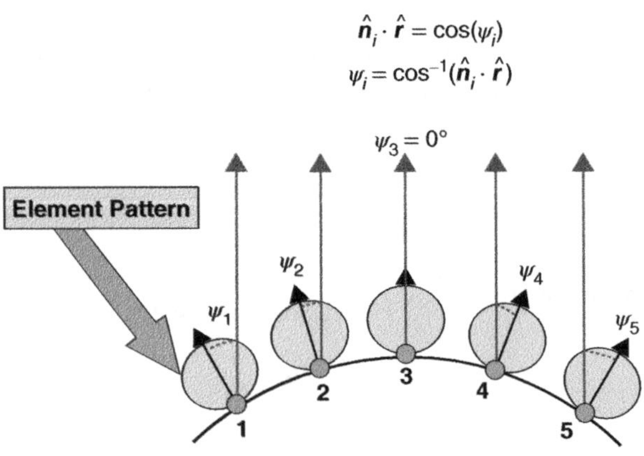

Figure 2.23 Different element pattern contributions for each individual element in a conformal array with no scan (boresight).

Figure 2.23. At boresight, each element is looking through a different point in its element pattern giving a different contribution than its neighbors.

2.5.3 Example

2.5.3.1 Conformal One-Dimensional Array

In the following example, the array pattern for a curved linear source in the xz-plane will be calculated using Equation 2.34. The elements are assumed to lie on the arc of a circle with arbitrary radius R. Figure 2.23 depicts the example array. For simplicity, the element factor, EF, is assumed to be 1, and the amplitude weights, a_i, are assumed to be uniform (i.e., $a_i = 1$). Table 2.1 shows the simplified expressions for the variables in Equation 2.34. It is important to note that the expression for the element unit normal vector, $\hat{\mathbf{n}}_i$, in Table 2.1 is only applicable for this example geometry. For other curvatures, the expression must be modified appropriately. Substituting the expressions in Table 2.1 into Equation 2.34 gives the following equation,

$$F(\mathbf{r}) = \sum_{i}^{N} cos(\psi_i) e^{jk[x_i(sin\theta - sin\theta_o) + z(cos\theta - cos\theta_o)]}, \tag{2.35}$$

where $cos(\psi_i)$ for this example is equal to

$$cos(\psi_i) = \hat{\mathbf{n}}_i \cdot \hat{\mathbf{r}} = \frac{x_i sin\theta + z_i cos\theta}{\sqrt{x_i^2 + z_i^2}}. \tag{2.36}$$

Using Equation 2.36, the array pattern can then be easily computed using Equation 2.35.

Table 2.1 Variable expressions for the array pattern equation for a curved line source in the xz-plane.

Variable	Simplified Expression
EF	1
a_i	1, for all i
$\hat{\mathbf{r}}$	$sin\theta\hat{\mathbf{x}} + cos\theta\hat{\mathbf{z}}$
$\hat{\mathbf{r}}_o$	$sin\theta_o\hat{\mathbf{x}} + cos\theta_o\hat{\mathbf{z}}$
$\hat{\mathbf{r}}_i$	$x_i\hat{\mathbf{x}} + z_i\hat{\mathbf{z}}$
$\hat{\mathbf{n}}_i$	$\frac{\hat{\mathbf{r}}_i}{\lvert\hat{\mathbf{r}}_i\rvert}$

2.6 2D AESA Pattern Formulation

In Section 2.5.1, expressions were derived for a one-dimensional (1D) AESA. The motivation for this is that a majority of the fundamental AESA concepts can be derived from the 1D expression. In practice, most AESAs are two-dimensional (2D) arrays; however, the theory expounded upon in Section 2.5.1 can be extended to the 2D case. Figure 2.24 shows an illustration of a 2D array of elements. The AESA antenna elements are positioned in the xy-plane and are assumed to radiate in the +z direction, or the forward hemisphere. As will be discussed later, this coordinate orientation is the same as what is traditionally called antenna coordinates. Each element is assumed to possess either a phase shifter or a time delay unit to electronically scan the beam. A manifold is assumed to be behind the elements summing their individual contributions coherently.

In Section 2.2.1, the element spacing was represented by *d*. In the 2D case, two element spacing values must now be specified. The spacing in the x dimension will be denoted by *dx*, and *dy* will denote the spacing in the y dimension. The number of elements in the x direction will be represented by *M* (identical to Section 2.2.1), and *N* will be used to represent the number of elements in the y direction. The total number of elements can then be expressed as $M \cdot N$. After defining representations for the element spacing and number of elements in the x and y dimensions, we can now express equations for the x-y element positions in the array as

$$x_m = (m - 0.5(M + 1))d_x, \text{ where } m = 1, \dots, M \tag{2.37}$$

$$y_n = (n - 0.5(N + 1))d_y, \text{ where } n = 1, \dots, N. \tag{2.38}$$

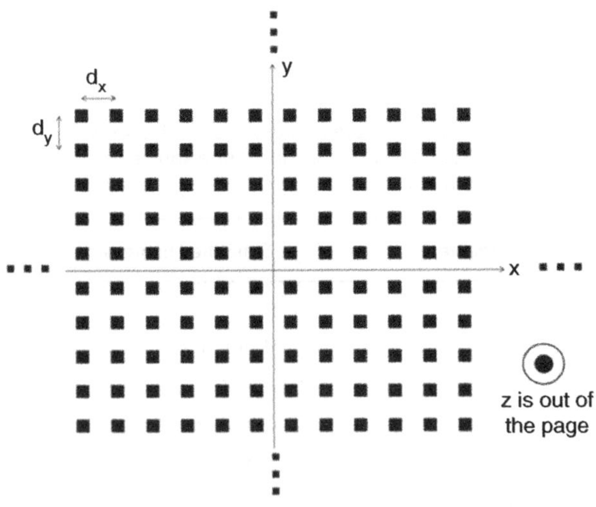

Figure 2.24 Two-dimensional AESA grid.

Equations 2.37 and 2.38 specify a rectangular grid of elements whose phase center is located at (0,0). The indexing used in Equations 2.37 and 2.38 is not unique in that the element spacing can be specified differently with a different phase center location.

The expression for a 1D *AF* was shown in Section 2.5.1 to be:

$$AF = \sum_{m=1}^{M} A_m e^{j(\frac{2\pi}{\lambda} x_m sin\theta)}. \quad (2.39)$$

This must now be expanded to include the additional elements in the y dimension. The 2D *AF* can be expressed as:

$$AF = \sum_{l=1}^{M \cdot N} C_l e^{j(\frac{2\pi}{\lambda} x_l sin\theta cos\phi + \frac{2\pi}{\lambda} y_l sin\theta sin\phi)}, \quad (2.40)$$

where C_l is a complex voltage that can be represented as $C_l = c_l e^{j\Theta_l}$. Setting $\Theta_l = (\frac{2\pi}{\lambda} x_l sin\theta_o cos\phi_o + \frac{2\pi}{\lambda} y_l sin\theta_o sin\phi_o)$, Equation 2.40 can then be expressed as:

$$AF = \sum_{l=1}^{M \cdot N} c_i e^{j[(\frac{2\pi}{\lambda} x_l sin\theta cos\phi + \frac{2\pi}{\lambda} y_l sin\theta sin\phi) - (\frac{2\pi}{\lambda} x_l sin\theta_o cos\phi_o + \frac{2\pi}{\lambda} y_l sin\theta_o sin\phi_o)]}. \quad (2.41)$$

Rearranging terms in Equation 2.41 and substituting $a_l \cdot b_l$ for c_l yields:

$$AF = \sum_{l=1}^{M \cdot N} a_l e^{j[(\frac{2\pi}{\lambda} x_l sin\theta cos\phi - \frac{2\pi}{\lambda} x_l sin\theta_o cos\phi_o)]} \cdot b_l e^{j[(\frac{2\pi}{\lambda} y_l sin\theta sin\phi - \frac{2\pi}{\lambda} y_l sin\theta_o sin\phi_o)]}. \quad (2.42)$$

If we assume that for each row *n* the a_l are constant and that for each column *m* the b_l are constant, Equation 2.42 can be written as:

$$AF = \sum_{m=1}^{M} a_m e^{j[(\frac{2\pi}{\lambda} x_m sin\theta cos\phi - \frac{2\pi}{\lambda} x_m sin\theta_o cos\phi_o)]} \cdot \sum_{n=1}^{N} b_n e^{j[(\frac{2\pi}{\lambda} y_n sin\theta sin\phi - \frac{2\pi}{\lambda} y_n sin\theta_o sin\phi_o)]}. \quad (2.43)$$

This condition is defined as separable weights, which means the 2D *AF* can be calculated by multiplying the 1D *AF*s for x and y. For weightings that are not separable, such as circular weighting, Equation 2.43 cannot be used, and Equation 2.41 should be used. Using Equation 2.41, the total 2D array pattern is

$$F(\theta, \phi) = cos^{\frac{EF}{2}} \theta \sum_{l=1}^{M \cdot N} c_i e^{j[(\frac{2\pi}{\lambda} x_l sin\theta cos\phi + \frac{2\pi}{\lambda} y_l sin\theta sin\phi) - (\frac{2\pi}{\lambda} x_l sin\theta_o cos\phi_o + \frac{2\pi}{\lambda} y_l sin\theta_o sin\phi_o)]}. \quad (2.44)$$

2.6.1 AESA Spatial Coordinate Definitions

When computing the spatial pattern for an AESA it is important to delineate what coordinate system is being used. Depending on the application, some coordinate systems may be more advantageous than others. Figure 2.25 depicts a 2D AESA in

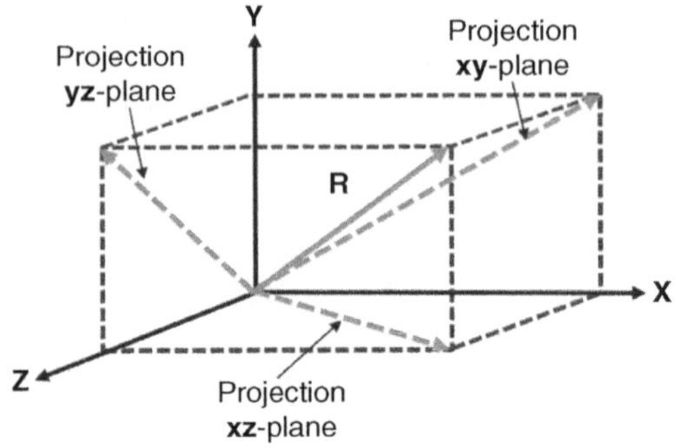

Figure 2.25 Coordinate system for a two-dimensional AESA in three-dimensional space. The AESA is assumed to be in the x-y plane. (Batzel et al., 2010).

three dimensional space. For convenience, the AESA is positioned in the xy-plane, and it is radiating in the +z direction. The +z direction is referred to as the forward hemisphere, and the −z direction is referred to as the backward hemisphere. The point R as shown in Figure 2.25 represents a point in space whose origin is at (0,0,0) and coincides with the boresight position of the AESA. The dashed lines, which reside in the xy-, yz-, and xz-planes, are the projections of point R in those planes. This picture will serve as the basis for understanding the other coordinate systems described in this section. In the following coordinate systems discussed, the purpose is to represent each point in space with a corresponding angle pair. These angles are then used to describe the spatial distribution of the AESA pattern, which correspondingly relates to performance.

2.6.2 Antenna Coordinates

Figure 2.26 depicts what is commonly called antenna coordinates. In this coordinate system each point R in space is represented by the angles θ_z and ϕ. θ_z is the angle subtended from the z-axis to the point R. The angle ϕ is the angle between the projection of R onto the xy-plane and the x-axis. This coordinate system definition is very intuitive as it is simply the spherical coordinate system. Correspondingly,

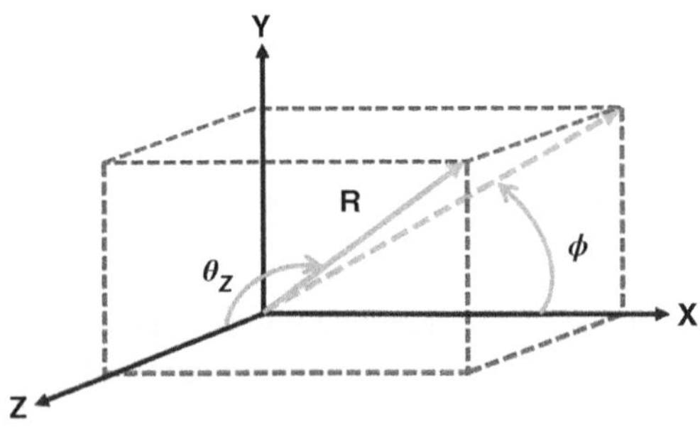

Figure 2.26 Antenna coordinates.

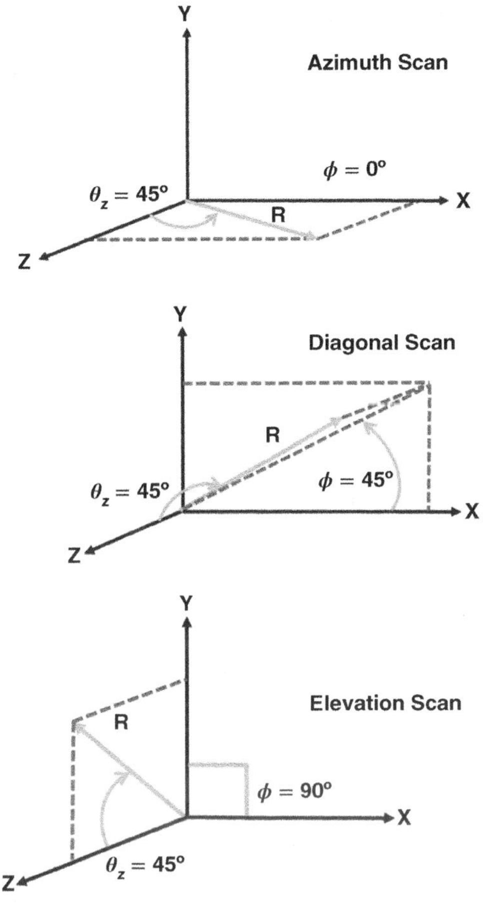

Figure 2.27 Antenna coordinates: constant $\theta_z = 45^\circ$ with $\phi = 0^\circ$ (azimuth scan), $\phi = 45^\circ$ (intercardinal scan), and $\phi = 90^\circ$ (elevation scan).

the point R can be represented as $R = (sin\theta_z cos\phi, sin\theta_z sin\phi, cos\theta_z)$ when the magnitude of R is set to unity. As an example, if an AESA's main beam is scanned in elevation to 45° there is a corresponding θ_z and ϕ that are associated with that point in space. For this example, this would correspond to $\theta_z = 45^\circ$, $\phi = 90^\circ$ as shown in Figure 2.27. Figure 2.27 also shows two different scan conditions when $\theta_z = 45^\circ$. The azimuth scan refers to scanning the beam in the xz-plane ($\phi = 0^\circ$), and the elevation scan refers to scanning the beam in yz-plane ($\phi = 90^\circ$).

Table 2.2 lists the values of θ_z and ϕ for the three different scan cases. When θ_z is kept constant, this traces a cone out in space whose apex is at z = 0 and whose base traces out a circle parallel to the xy-plane. This is typically called a scan cone angle. If we were to rotate Figure 2.27 about the z-axis and view it looking perpendicular to the xy-plane, the lines would trace out a circle.

Table 2.2 Antenna coordinate angle values for various scan types.

Scan type	θ_i	ϕ_i
Azimuth scan	$\theta_z = 0°$ to $90°$	$\phi = 0°,\ 180°$
Intercardinal scan	$\theta_z = 0°$ to $90°$	$\phi = 45°,\ 135°,\ 225°,\ 315°$
Elevation scan	$\theta_z = 0°$ to $90°$	$\phi = 90°,\ 270°$

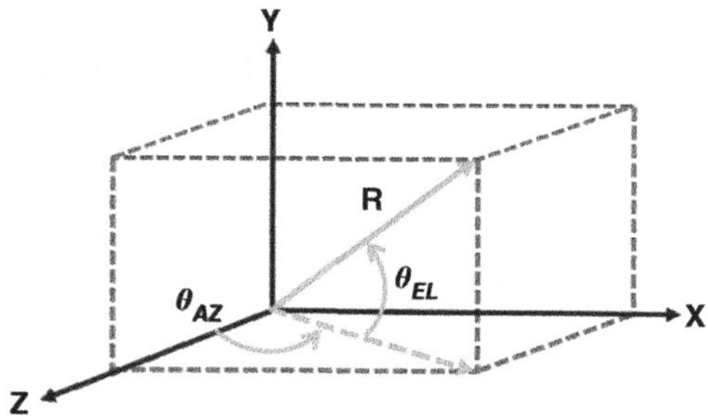

Figure 2.28 Radar coordinates (Batzel et al., 2010).

2.6.3 Radar Coordinates

Figure 2.28 graphically defines the spatial angles used for radar coordinates. Similar to antenna coordinates, two angles are required to define a point in three-dimensional space. The two angles are θ_{AZ} and θ_{EL}. θ_{AZ} is defined as the angle subtended by the projection of R onto the xz-plane with the z-axis. θ_{EL} is defined as the angle subtended by the vector to point R and the xz-plane. We'll use the same example as in the previous section and observe what values of θ_{AZ} and θ_{EL} correspond to a 45° scan in elevation. Figure 2.29 shows that for this scan case, $\theta_{AZ} = 0°$ and $\theta_{EL} = 45°$. From a radar perspective, this coordinate system is more intuitive than that of antenna coordinates. In a radar system, the AESA

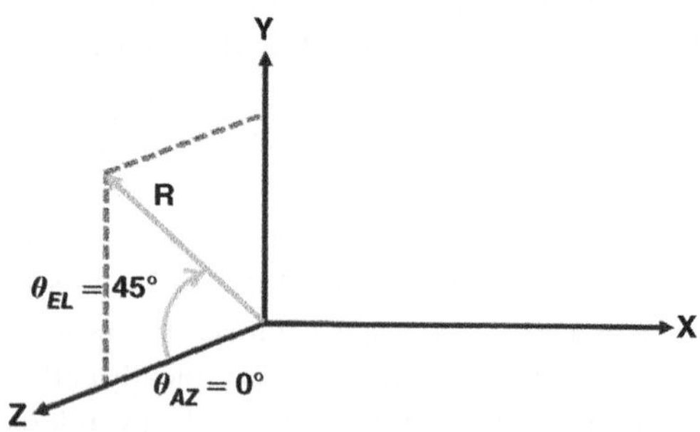

Figure 2.29 Radar coordinates: elevation scan to 45°.

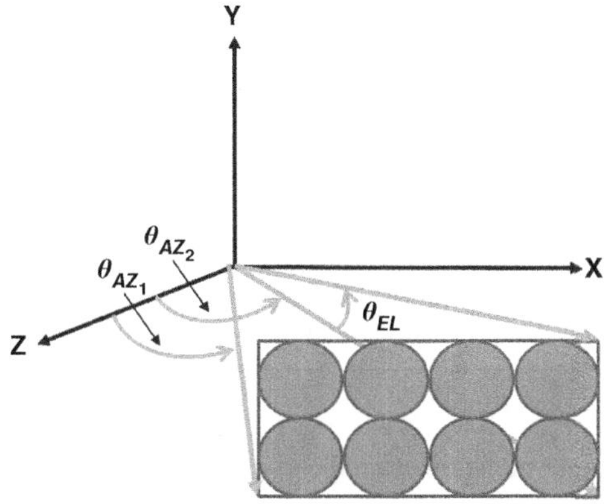

Figure 2.30 Raster scan of AESA beams (Batzel et al., 2010).

main beam is typically scanned in some type of raster fashion where the beams are distributed spatially in rows and columns. Figure 2.30 shows an exemplar of a raster scan of a beam in azimuth and elevation. In the figure each row corresponds to an azimuth scan where θ_{EL} is constant and θ_{AZ} is varied. This type of application lends itself to the radar coordinate system.

2.6.4 Antenna Cone Angle Coordinates

Figure 2.31 shows a pictorial definition of antenna cone angle coordinates. The two angles that specify a point in space for this coordinate system are θ_A and θ_E. θ_A is defined as the angle subtended by the projection of R onto the yz-plane and R. θ_E is defined as the angle between the point R and the xz-plane. From this definition we see that $\theta_E = \theta_{EL}$ (radar coordinates).

Any of the coordinate systems previously mentioned can be used to calculate spatial (i.e., Cartesian) coordinates for an AESA. In a real system application, an antenna engineer may be using a coordinate system different than the system engineer. Because of this, it is good to have the angular transformations between coordinates systems so that there is consistency in requirements flowdown and system

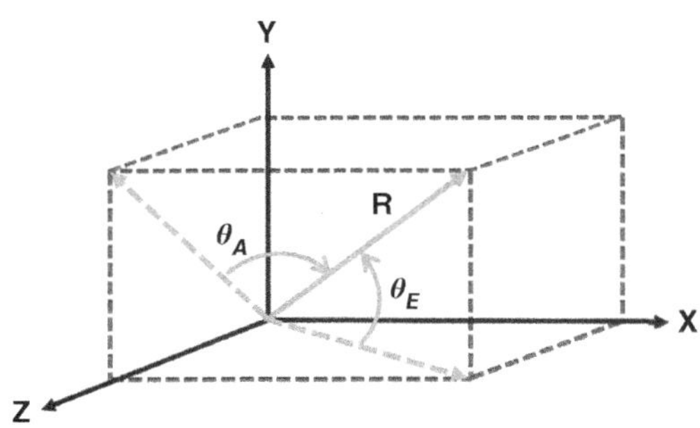

Figure 2.31 Antenna cone angle coordinates (Batzel et al., 2010).

Table 2.3 Angle transformation given antenna coordinate system angles.

Given Antenna Angles θ_z and ϕ		
Radar coordinates	θ_{AZ}	$tan^{-1}\left(\frac{sin\theta_z \cdot cos\phi}{cos\theta_z}\right)$
	θ_{EL}	$sin^{-1}(sin\theta_z \cdot sin\phi)$
Antenna cone angles	θ_A	$sin^{-1}(sin\theta_z \cdot cos\phi)$
	θ_E	$sin^{-1}(sin\theta_z \cdot sin\phi)$

Table 2.4 Angle transformation given radar coordinate system angles.

Given Radar Angles θ_{AZ} and θ_{EL}		
Antenna angles	θ_z	$cos^{-1}(cos\theta_{AZ} \cdot cos\theta_{EL})$
	ϕ	$tan^{-1}\left(\frac{sin\theta_{EL}}{sin\theta_{AZ} \cdot cos\theta_{EL}}\right)$
Antenna cone angles	θ_A	$sin^{-1}(sin\theta_{AZ} \cdot cos\theta_{EL})$
	θ_E	θ_{EL}

Table 2.5 Angle transformation given antenna cone angle coordinate system angles.

Given Antenna Cone Angles θ_A and θ_E		
Antenna angles	θ_z	$sin^{-1}(\sqrt{sin^2(\theta_A) - sin^2(\theta_E)})$
	ϕ	$tan^{-1}(sin\theta_E / sin\theta_A)$
Radar angles	θ_{AZ}	$sin^{-1}\left(\frac{sin\theta_A}{cos\theta E}\right)$
	θ_{EL}	θ_E

performance evaluation. Tables 2.3–2.5 provide a summary of the angular transformations for the antenna, radar, and antenna cone angle coordinate systems.

2.6.5 Sine Space Representation

An alternative to using angular coordinates for modeling AESAs is the sine space representation. Sine space is simply a hemispherical projection of

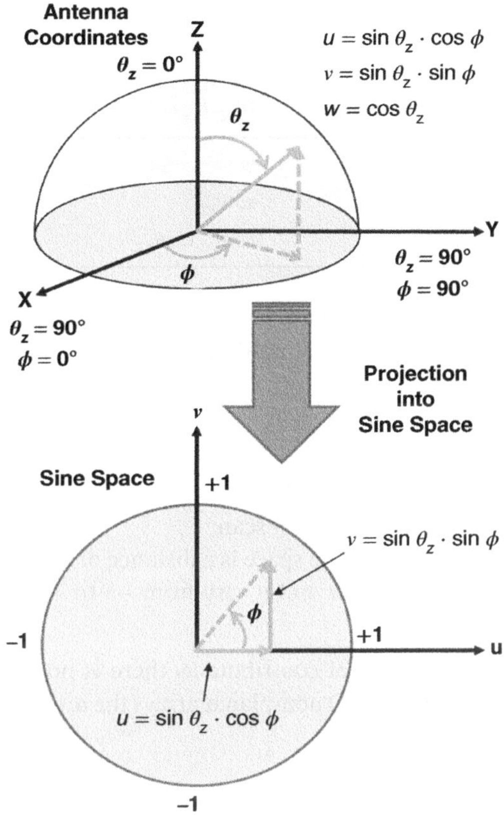

Figure 2.32 Sine space representation (Long and Schmidt, 2010).

three-dimensional space onto a two-dimensional surface. Figure 2.32 graphically illustrates how three-dimensional space is mapped into two-dimensional space. Sine space is represented by the following variables: u, v, and w. Although the sine space variables can be computed using any of the three angular coordinate systems previously discussed, antenna coordinates provide a very intuitive comparison and will be elaborated on in the following discussion. The conversion to sine space from the other coordinates systems is shown in Table 2.6. The expressions for sine space are

$$u = sin\theta_z cos\phi, \tag{2.45}$$

$$v = sin\theta_z sin\phi, \tag{2.46}$$

$$w = cos\theta_z. \tag{2.47}$$

Table 2.6 Conversion to sine space from angular coordinates.

Sine Space	**Antenna** (θ_z, ϕ)	**Radar** $(\theta_{AZ}, \theta_{EL})$	**Antenna Cone** (θ_A, θ_E)
u	$sin\theta_z cos\phi$	$sin\theta_{AZ} cos\theta_{EL}$	$sin\theta_A$
v	$sin\theta_z sin\phi$	$sin\theta_{EL}$	$sin\theta_E$
w	$cos\theta_z$	$cos\theta_{AZ} cos\theta_{EL}$	$cos(sin^{-1}(\frac{sin\theta_A}{cos\theta_E}))cos\theta_E$

These expressions are the traditional expressions for x, y, and z in spherical coordinates. Using Equations 2.45–2.47, a simplified form for the 2D *AF* can be written as

$$AF = \sum_{l=1}^{M\cdot N} c_l e^{j[(\frac{2\pi}{\lambda}x_l u + \frac{2\pi}{\lambda}y_l v) - (\frac{2\pi}{\lambda}x_l u_o + \frac{2\pi}{\lambda}y_l v_o)]}. \tag{2.48}$$

Several characteristics of the *AF* in sine space are:

- Constant beamwidth independent of scan;
- Peak of the scanned beam in sine space is a distance of $sin\theta_z$ from (0,0);
- For the forward hemisphere, u and v vary from -1 to $+1$ and, w varies from 0 to 1.

For a planar array, w does not contribute as there is no z component in the exponent of Equation 2.48. For 2D non-planar arrays the w term must be included.

2.6.6 AESA Element Grid

An AESA is composed of antenna elements that are arranged according to a defined element spacing. Previously, a linearly spaced 1D array of elements was introduced. It was shown that grating lobes arise due to the periodic nature of the *AF* and are a function of the element spacing. For a 2D configuration, the same relationships holds between grating lobes and element spacing with the difference being there are grating lobes in both the x and y dimensions spatially. An alternative to a rectangular grid of elements is a triangular grid. The triangular grid has unique properties, which will be elaborated on further in this chapter.

2.6.6.1 Rectangular Grid

Figure 2.33 illustrates a rectangular grid of elements with a linear spacing in both the x and y dimensions. Similar to the 1D case, now an expression is required to determine the grating lobes that occur due to the arrangement of elements in the y dimension. The equations for the grating lobes can be represented as (Mailloux, 1993):

$$\begin{aligned} u_m &= u_o + m\frac{\lambda}{d_x}, \quad m = 0, \pm 1, \pm 2, \ldots \\ v_n &= v_o + n\frac{\lambda}{d_y}, \quad n = 0, \pm 1, \pm 2, \ldots \end{aligned} \tag{2.49}$$

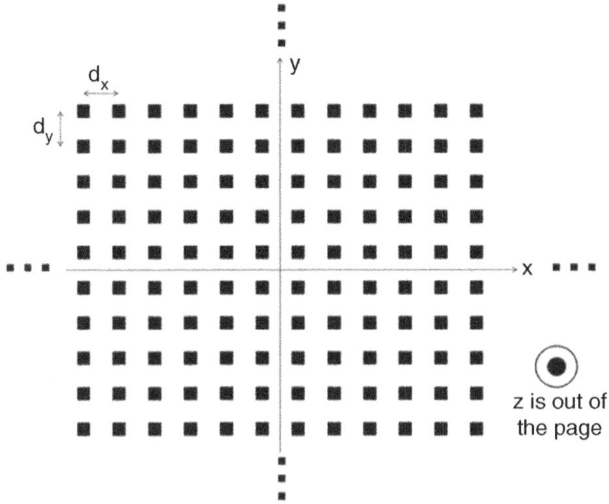

Figure 2.33 A two-dimensional AESA rectangular grid.

with the following relations:

$$\begin{aligned} cos\theta_{mn} &= (1 - u_m^2 - v_n^2)^{\frac{1}{2}}, \\ u_m^2 + v_n^2 &\leq 1. \end{aligned} \tag{2.50}$$

Figure 2.34 shows the grating lobe locations in sine space for a 2D rectangular grid with spacing d_x and d_y. In Figure 2.34, the region inside the unit circle is referred to as visible space and is shaded. This is because the constant circle of radius 1 corresponds to a constant circle of $\theta = 90°$, which is the extent of the forward hemisphere. In Equation 2.49, u_o corresponds to the main beam and for boresite ($\theta_o = 0°$ and $\phi_o = 0°$) has a value of 0. What is readily apparent is that as the main beam is electronically scanned, the grating lobe locations move with the main beam with a fixed offset that is proportional to an integer multiple of $\frac{\lambda}{d}$. This is demonstrated in Figures 2.35–2.37. Each figure shows the grating lobe locations for a boresite condition with the overlaid position of the AESA beam and its corresponding grating lobes. The element spacings shown are for $d_x = d_y = \frac{\lambda}{4}, \frac{\lambda}{2}$, and λ. For all cases, when the main beam is electronically scanned, the grating lobes move with the main beam. If the element spacing is greater than half wavelength, for certain angles grating lobes will be present in visible space, which is undesired and unacceptable for most applications. In contrast, if the element spacings are made less than half wavelength, more margin is added, albeit at a price. More electronics are needed for the increased number of elements, which translates to a more expensive AESA.

The region in visible space where the main beam can be scanned is referred to as the grating lobe free scan volume. This can simply be illustrated by drawing unit circles around each of the grating lobes. Wherever the grating lobe circles

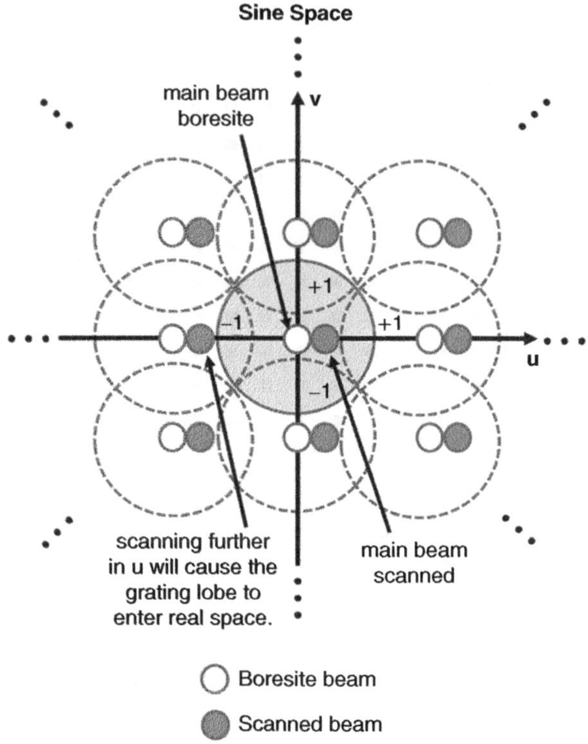

Figure 2.34 Grating lobe diagram for a rectangular grid with half-lambda spacing (Davis, 2010).

do not intersect visible space corresponds to the grating lobe free scan volume. Figures 2.38 and 2.39 illustrate the dependence of the grating lobe free scan volume on the element spacing. In Figure 2.38, the grating lobe free scan volume is the entire unit circle or all of visible space. The AESA beam can be scanned electronically anywhere in visible space without the presence of grating lobes. In Figure 2.39 this is not the case. The grating lobe free scan volume for the larger element spacing is now limited to a portion of visible space. For scan angles corresponding to the overlapped unit circles of the main beam and its associated grating lobes, grating lobes will appear in real space. Only in the regions where there is no overlap of the unit circles can the AESA beam be scanned without grating lobes. This is denoted by the shaded region in Figure 2.39. An additional grating lobe characteristic is that in the intercardinal (diagonal) plane the AESA can be scanned farther. This is because the diagonal distance of the main beam from the diagonal grating lobes is $\sqrt{2} \cdot \frac{\lambda}{d}$. For applications where the grating lobe free scan volume in Figure 2.39 would be acceptable, a cost savings would be realized in electronics because half-wavelength spacing is not required, which means a reduced element count.

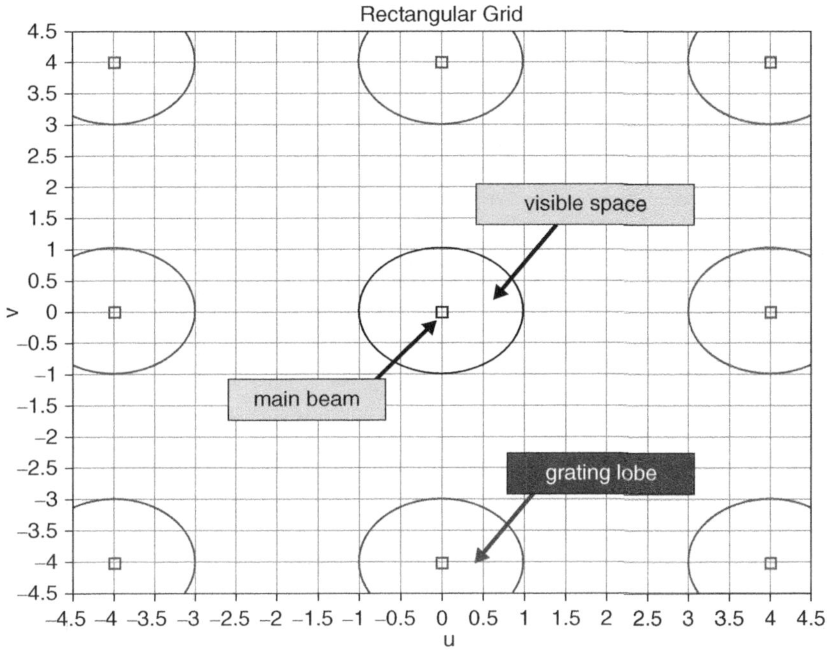

Figure 2.35 Grating lobe diagram for a rectangular grid for boresite and electronic scan ($d_x = d_y = \frac{\lambda}{4}$).

2.6.6.2 Triangular Grid

Figure 2.40 shows a triangular grid of elements. In the previous section, it was mentioned that a way to save cost is by reducing the number of elements using element spacing greater than half wavelength. Triangular grids provide another way to reduce element count while maintaining scan performance. For a rectangular grid, the area per element is $d_x \cdot d_y$. The area per element for a triangular grid is $2d_x \cdot d_y$. For a fixed aperture size, less elements are required for a triangular grid. Furthermore, it can be shown that for the same amount of grating lobe suppression, a rectangular grid requires 16% more elements than a triangular grid (Skolnik, 1990).

Using the element spacing definitions as shown in Figure 2.40, the expressions for the grating lobes are (Skolnik, 1990):

$$\begin{aligned} &u_m = u_o + m\frac{\lambda}{2d_x}, \quad v_n = v_o + n\frac{\lambda}{2d_y}, \\ &m, n = 0, \pm 1, \pm 2, \dots \\ &m + n \text{ is even.} \end{aligned} \tag{2.51}$$

The derivation for these expressions is included in Appendix C. Using the expressions in Equation 2.51, grating lobe free scan volume plots can be shown similar to

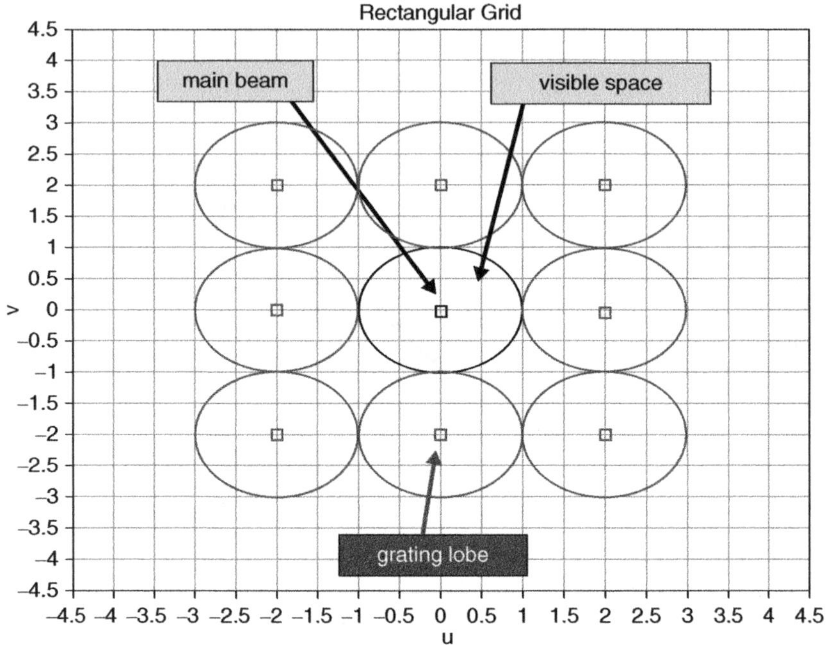

Figure 2.36 Grating lobe diagram for a rectangular grid for boresite and electronic scan ($d_x = d_y = \frac{\lambda}{2}$).

those in the previous section. Figure 2.41 shows the grating lobe free scan volume for a triangular grid. In Figure 2.41 we see that in sine space the grating lobes are oriented in a triangular manner just as the element grid. This can be advantageous in applications where the triangular grid provides a better scan volume due to the orientation of the grating lobes.

2.6.7 Two-Dimensional Pattern Synthesis

The expression for a 2D AESA array factor was shown to be

$$AF = \sum_{l=1}^{M\cdot N} c_l e^{j[(\frac{2\pi}{\lambda}x_l u + \frac{2\pi}{\lambda}y_l v) - (\frac{2\pi}{\lambda}x_l u_o + \frac{2\pi}{\lambda}y_l v_o)]}. \tag{2.52}$$

The complete expression for the 2D AESA pattern is therefore

$$F(\theta, \phi) = \cos^{\frac{EF}{2}}\theta \cdot \sum_{l=1}^{M\cdot N} c_l e^{j[(\frac{2\pi}{\lambda}x_l u + \frac{2\pi}{\lambda}y_l v) - (\frac{2\pi}{\lambda}x_l u_o + \frac{2\pi}{\lambda}y_l v_o)]}. \tag{2.53}$$

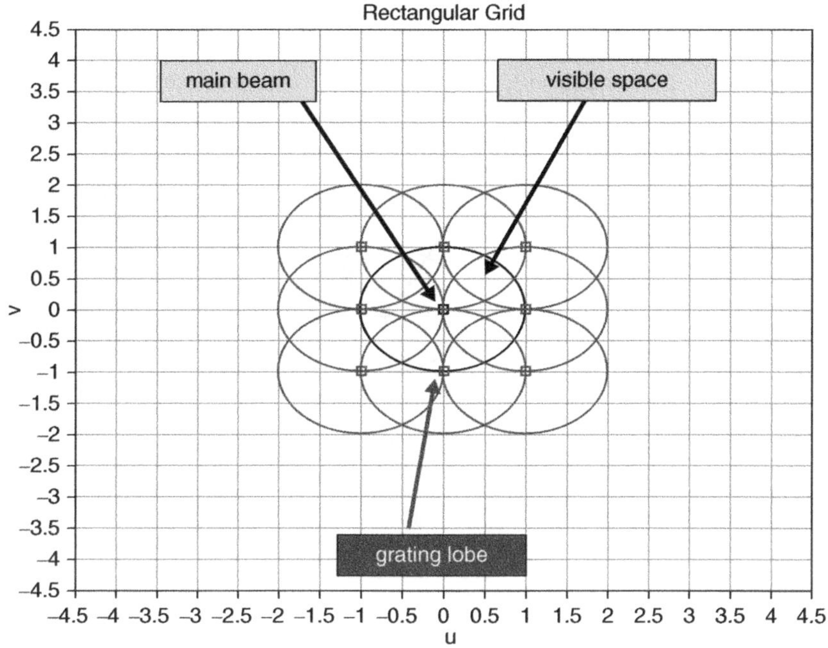

Figure 2.37 Grating lobe diagram for a rectangular grid for boresite and electronic scan ($d_x = d_y = \lambda$).

The remainder of this chapter will illustrate 2D array patterns plotted in sine space for convenience. The angular equivalent plots can easily be generated by using the conversions from sine space to angle space shown in Table 2.6. The following three subsections will show patterns for a 2D AESA. An element spacing less than 0.5 λ is used to show grating lobes effects for electronic scan beyond 60°.

2.6.7.1 Ideal Patterns

Ideal patterns using Equation 2.53 are useful to provide a high level of confidence in the AESA's scan performance. Although it is necessary to include the impacts of errors, using ideal patterns provides an initial basis to start within a design. Amplitude and phase errors perturb the ideal pattern and primarily affect sidelobe levels. A great benefit of AESAs that have a large number of elements is that the main beam remains relatively unaltered. The main beam of an AESA is well behaved even in the presence of errors.

Figure 2.42 and Figure 2.43 show boresite antenna patterns in radar coordinates and sine space, respectively. The remaining plots will be shown in sine space. The plots in Figure 2.44 show the antenna pattern as a function of electronic scan in the principal planes and in the intercardinal (diagonal) plane. Figure 2.45 shows

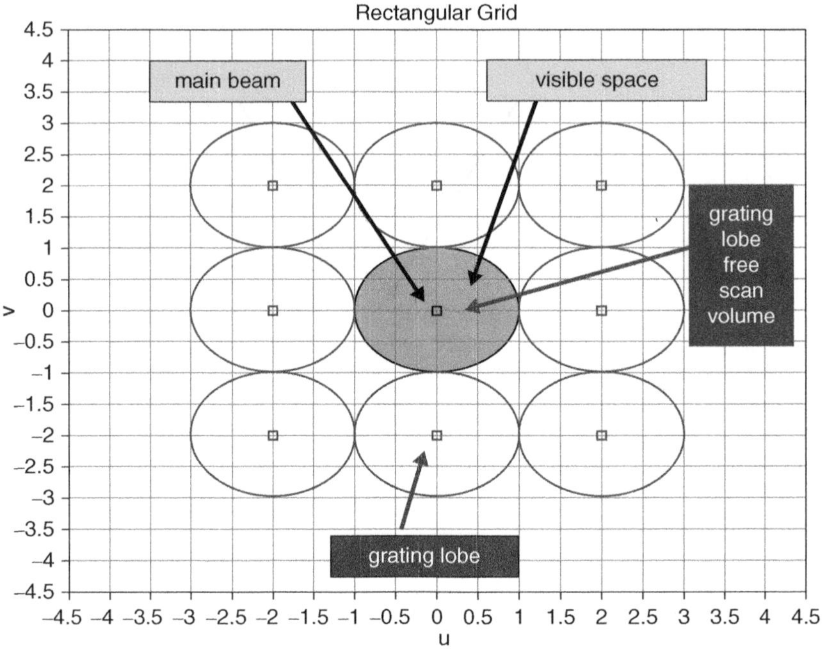

Figure 2.38 Grating lobe free scan volume for a max scan of 90° in both azimuth and elevation for $d_x = d_y = \frac{\lambda}{2}$.

the array pattern for electronic scan beyond 60°. The element spacing chosen for this example does not support a grating lobe free scan volume for scan angles greater than 60°. Thus, the grating lobes appear in real space, which is undesired. In the 2D plots, the grating lobe can be seen coming into real space. To avoid this in practice, margin can be added to the element spacing equations shown in Chapter 1.

Figure 2.46 highlights the reduction in SLL with uniform illumination and a 30 dB Taylor illumination. In applications where low SLLs are required, amplitude weighting is a valuable design tool. When amplitude weighting is implemented, there is no free lunch. A reduction in SLLs comes at the price of increased beamwidth and reduced gain. The reduction in gain is called the taper loss (Mailloux, 1993) and can be represented by

$$TL = \frac{|\sum c_l|^2}{n \sum |c_l|^2}, \tag{2.54}$$

where n is the number of elements, and c_l are the amplitude weights shown in Equation 2.53.

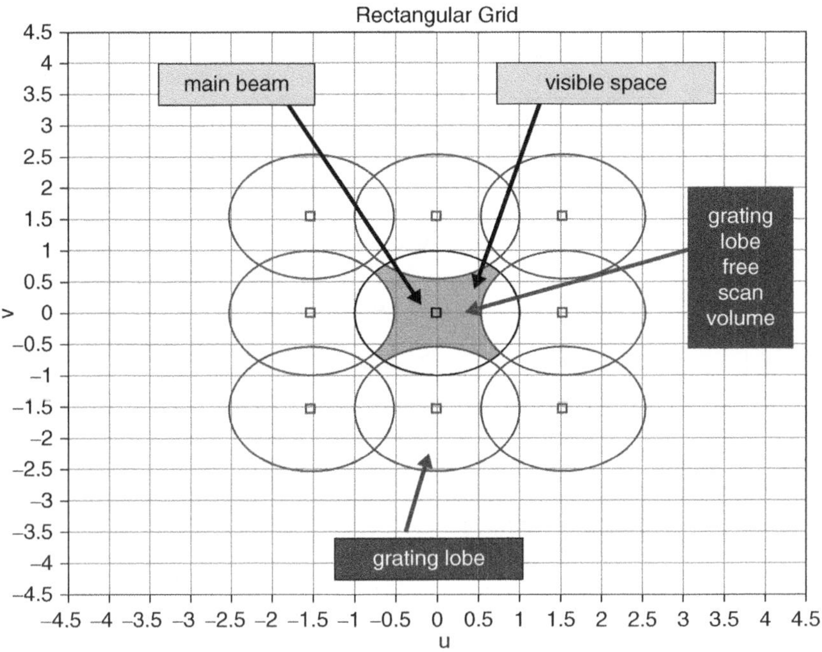

Figure 2.39 Grating lobe free scan volume for a max scan of 60° in both azimuth and elevation for $d_x = d_y = \frac{\lambda}{1.866}$.

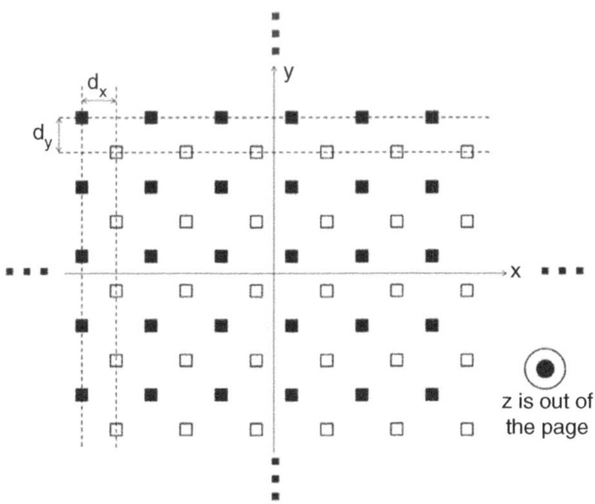

Figure 2.40 Triangular grid of array elements.

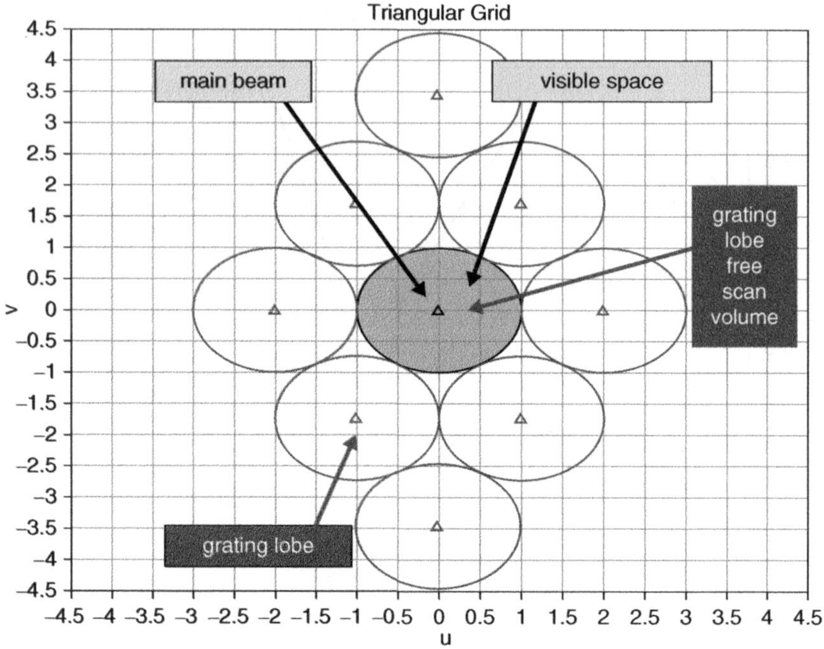

Figure 2.41 Grating lobe free scan volume for a max scan of 60° in both azimuth and elevation for $d_x = d_y = \frac{\lambda}{1.866}$ for a triangular grid.

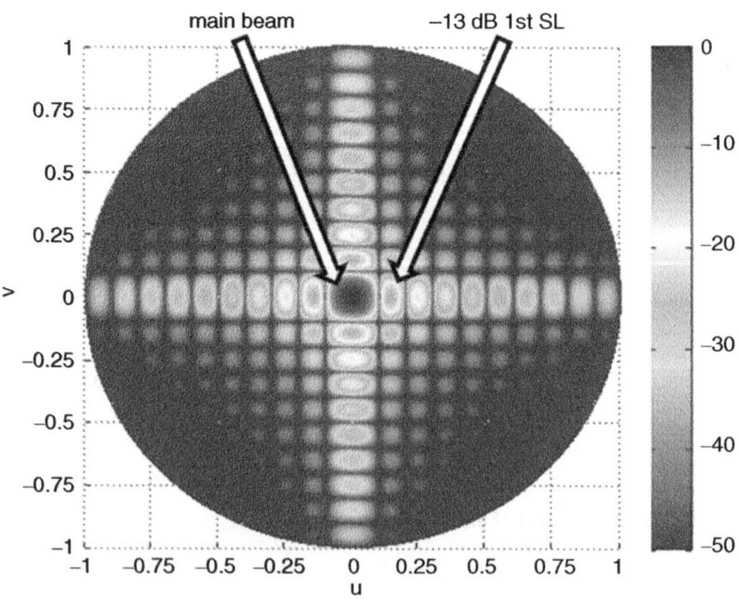

Figure 2.42 Boresite antenna pattern (no electronic scan) in radar coordinates.

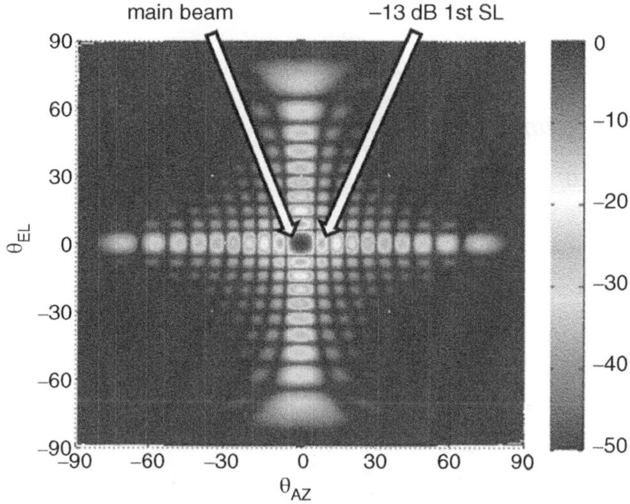

Figure 2.43 Boresite antenna pattern (no electronic scan) in sine space.

2.7 Circular Grid AESA Patterns

Figure 2.47 shows an AESA with a circular grid. The array is broken into annular rings having a constant radius for each ring denoted as r_{k,p_k}, where k denotes a particular ring, and p_k is the element number for the kth ring. As an example, in Figure 2.47, the first ring closest to the origin represents $k = 1$, and the furthest ring represents $k = 4$. Each of the elements for a specific ring has a constant angular spacing. The angle $\Delta\phi_k$ is the angle corresponding to the constant angular spacing for the kth ring.

Although the grid is different in structure than the rectangular grid, the formulation for the AESA pattern follows the same fundamental formula found in Equation 2.53. The difference is that the product $M \cdot N$ can be replaced with $M_{circular}$, which represent the total number of elements. This is shown in Equation 2.55.

$$F(\theta,\phi) = cos^{\frac{EF}{2}}\theta \cdot \sum_{l=1}^{M_{circular}} c_l e^{j[(\frac{2\pi}{\lambda}x_l u + \frac{2\pi}{\lambda}y_l v) - (\frac{2\pi}{\lambda}x_l u_o + \frac{2\pi}{\lambda}y_l v_o)]}. \tag{2.55}$$

Equation 2.55 can be rewritten to reflect the definitions for $r_{k,p}$ and $\Delta\phi_k$. Fundamentally, the total pattern is a sum of the contributions from each ring and can be expressed as

$$F(\theta,\phi) = cos^{\frac{EF}{2}}\theta \cdot \sum_{k=1}^{K}\sum_{p_k=1}^{P_k} c_{k,p_k} e^{j[(\frac{2\pi}{\lambda}x_{k,p_k} u + \frac{2\pi}{\lambda}y_{k,p_k} v) - (\frac{2\pi}{\lambda}x_{k,p_k} u_o + \frac{2\pi}{\lambda}y_{k,p_k} v_o)]}, \tag{2.56}$$

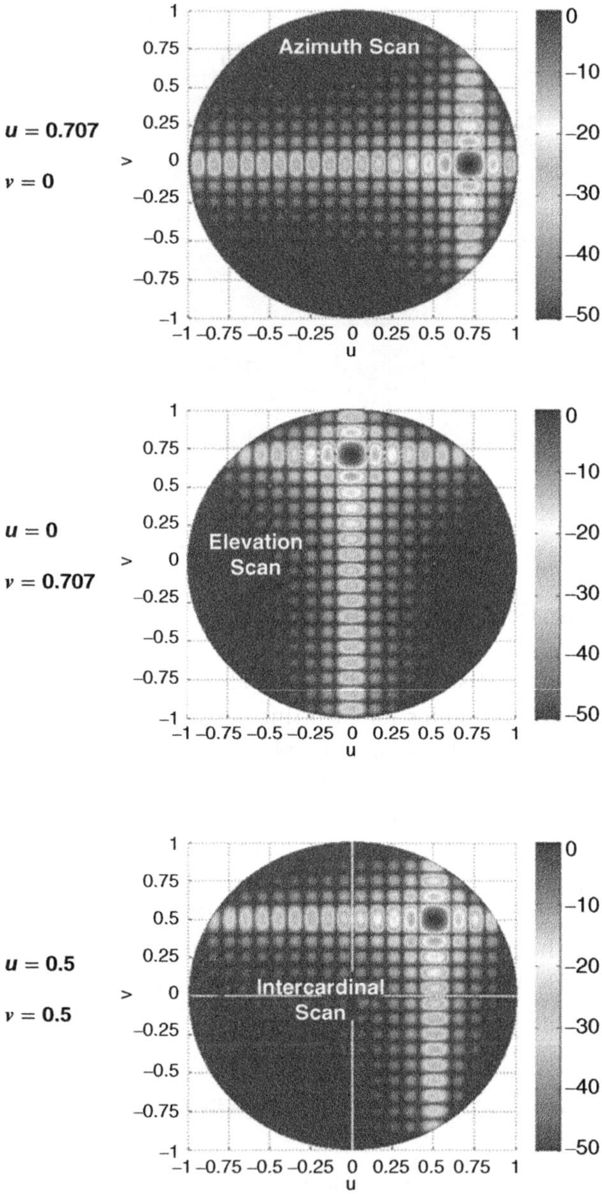

Figure 2.44 Electronically scanned antenna patterns in the principal planes and intercardinal plane.

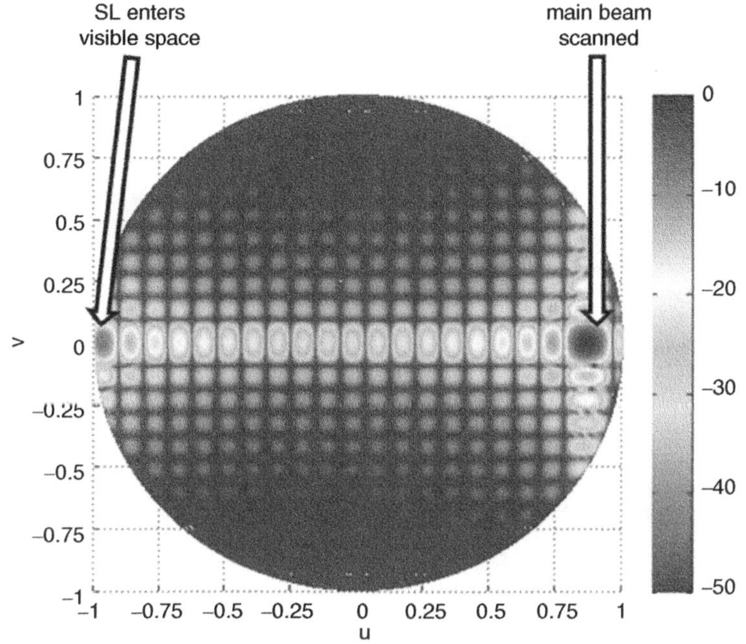

Figure 2.45 Electronic scan beyond 60° allows the fully formed grating lobe to appear in real space.

where $x_{k,p_k} = k\ r_{k,p_k} cos\left((p_k - 1)\Delta\phi_k\right)$ and $y_{k,p_k} = k\ r_{k,p_k} sin\left((p_k - 1)\Delta\phi_k\right)$. r_{k,p_k} and $\Delta\phi_k$ can be changed to maintain an average element spacing approximating $\frac{\lambda}{1+sin\theta}$. This has been done for the circular grid in Figure 2.47.

Figure 2.48 shows a 2D pattern for the circular grid in Figure 2.47. The SL structure for the circular grid is different than that for the rectangular grid. Instead of two sidelobe ridges in the principal azimuth and elevation planes, the SLs for the circular grid are annular in shape. Beyond the first annular ring, a majority of the SLs are 25 dB or lower. Additionally, unlike the uniform rectangular array that has a first SL at −13 dB, the first SL for a uniform circular array occurs at −17 dB. This is also depicted in Figure 2.49, which shows the azimuth pattern cut for Figure 2.48. Figure 2.50 shows a scanned pattern for the circular grid. Similar to the rectangular case, the SLs are translated spatially with the main beam. In this case of 60° electronic scan, the distinct 2D structure of the SLs is shown. Looking at the azimuth pattern cut in Figure 2.51, the pattern is relative well behaved. However, in Figure 2.50 the sidelobe ridges in the azimuth pattern cut extend spatially unlike the rectangular grid whose SL ridge are primarily contained in two planes. This can be mitigated by applying amplitude weighting to the circular grid array elements.

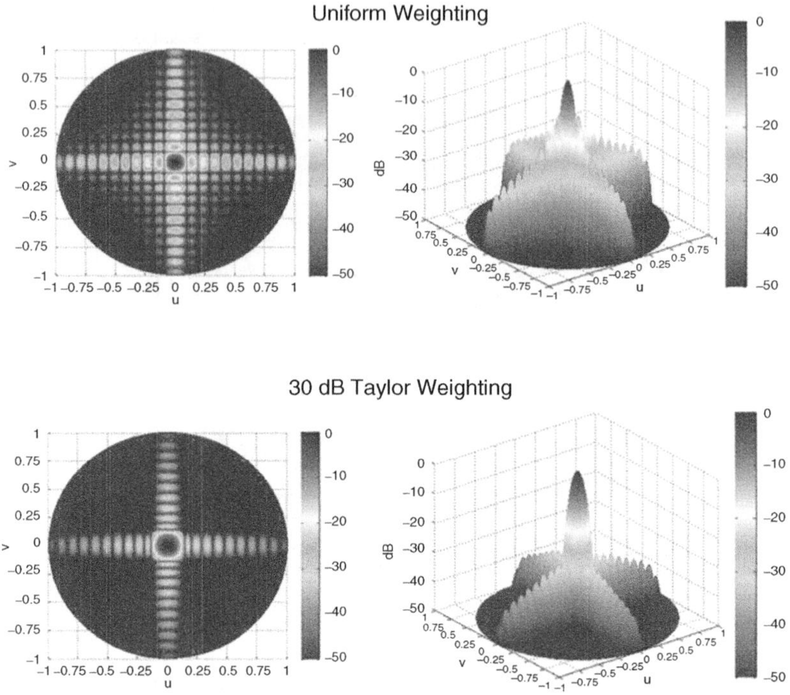

Figure 2.46 SLL reduction using a 30 dB Taylor weighting compared to a uniform distribution.

Figure 2.52 is an amplitude weighted pattern for the circular grid in Figure 2.47. The azimuth pattern cut is also shown in Figure 2.53. Similar to the rectangular grid, by employing amplitude weighting the SLs can be reduced. Using a circular grid is an effective approach for achieving maximum gain with low SLs, although a circular grid in practice is more challenging to manufacture. This is because combining or distributing signals is challenging from a mechanical perspective for the beamforming. A rectangular grid lends itself to easier manufacturing of the AESA in addition to repeatability. Also, the circular grid is not attractive in terms of scalability. For a rectangular grid, any rectangular-sized array can be built from rectangular building blocks. For a circular array, this is not true. Another way to achieve a circular grid pattern is to use a rectangular grid inscribed inside of a circle and remove elements outside of the circle. What is left is a hexagonal type of grid with rectangular spacing whose antenna pattern has a circular SL structure.

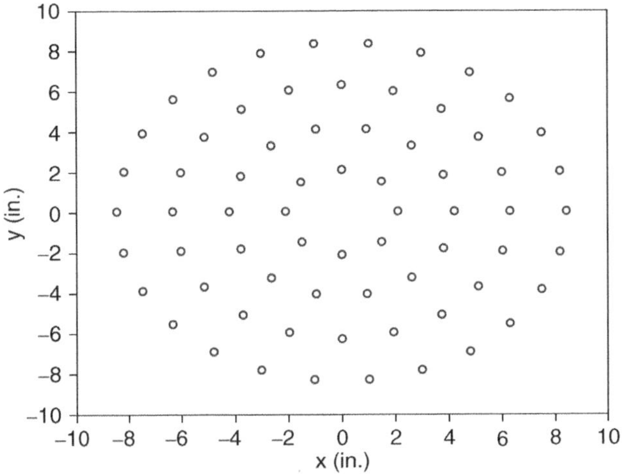

Figure 2.47 AESA with a circular grid. Each ring of the grid is separated by $\sim \frac{\lambda}{2}$. The radial distance between elements is $\sim \frac{\lambda}{2}$ also.

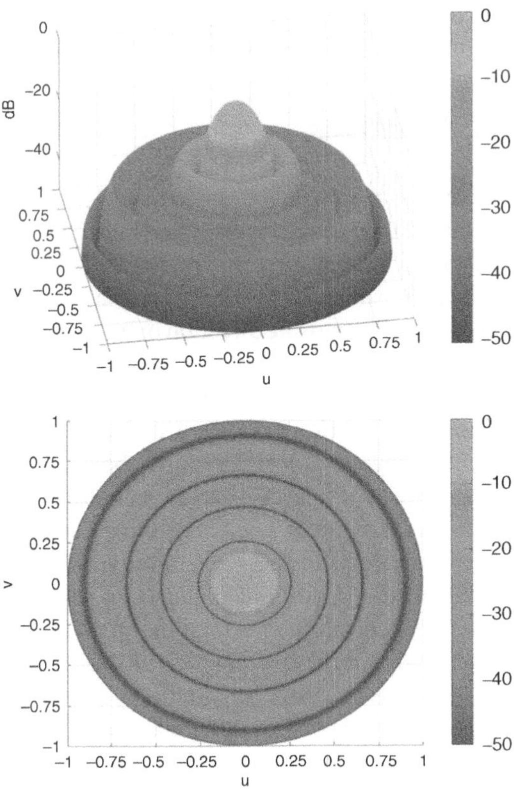

Figure 2.48 The circular grid in Figure 2.47 produces a pattern with grating lobes that are in annular rings. Although the first SL is lower than a rectangular grid AESA of the same size, the SL extends 360°.

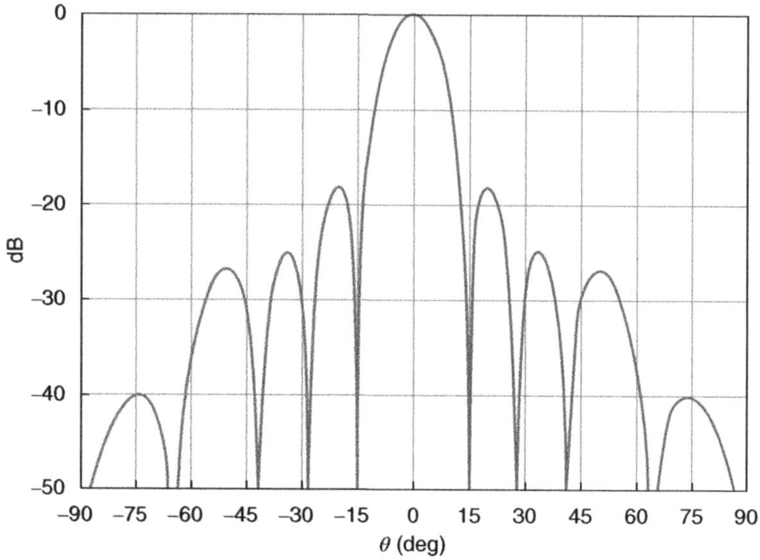

Figure 2.49 Azimuth pattern cut illustrates that a uniform circular array has a first SL at ~ −17 dB. This is 4 dB less than a uniform rectangular array.

2.8 Tilted AESA Patterns

Up to this point, we have assumed that the platform on which the AESA is installed is co-boresited with the AESA. In many installations this is not the case. The AESA may be installed in such a way that the normal of the AESA does not coincide with the normal of the platform. In Section 2.6.1, antenna coordinate systems were defined. In many instances, the boresite direction of the radar coordinates does not coincide with that of the AESA. As an example, radar AESAs for ship installations may be tilted upward to provide optimal coverage for the area above the horizon (the ocean). Additionally, on platforms that are subject to motion, such as a ship in rough seas, the AESA may be subject to motion about the x-, y-, and/or z-axis, which will change how the pattern is distributed spatially (Konapelsky, 2002).

Figure 2.54 shows a side view of an AESA tilted relative to the x-axis. In Figure 2.54 the x-axis is normal to the page. Different definitions can be used to describe rotation about the three principal axes. In this book we will use the following definitions. A pitch refers to a rotation about the x-axis, a roll refers to rotation about the z-axis (normal to the face of the array), and a yaw refers to a rotation about the y-axis. Figure 2.55 pictorially shows roll, pitch, and yaw as previously defined. For each of the different rotations, a transformation is required from the coordinate system of the array to that of the system. These

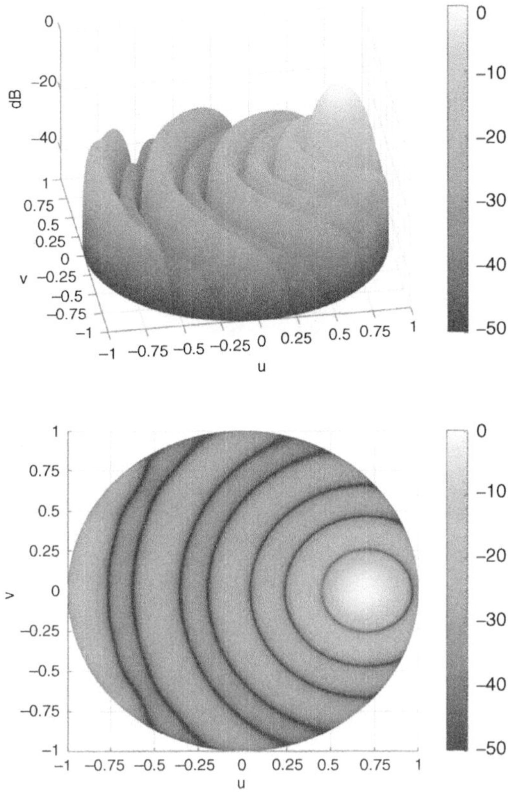

Figure 2.50 Circular grid pattern for a scan of 60°.

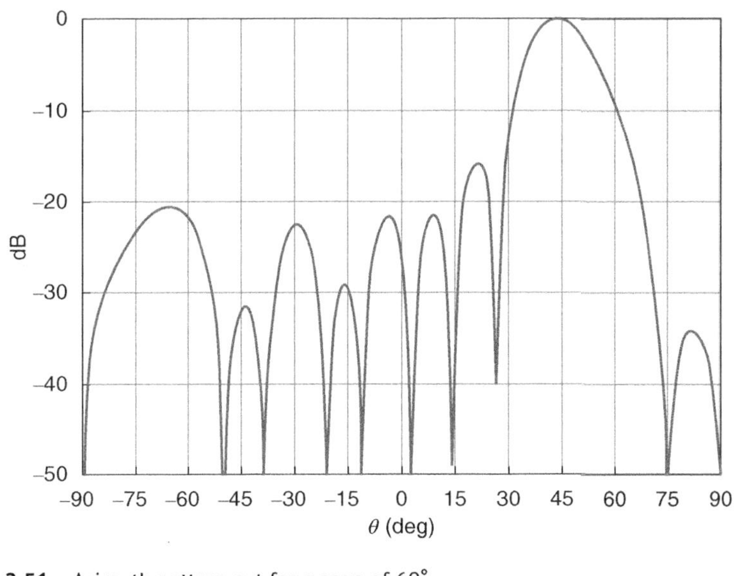

Figure 2.51 Azimuth pattern cut for a scan of 60°.

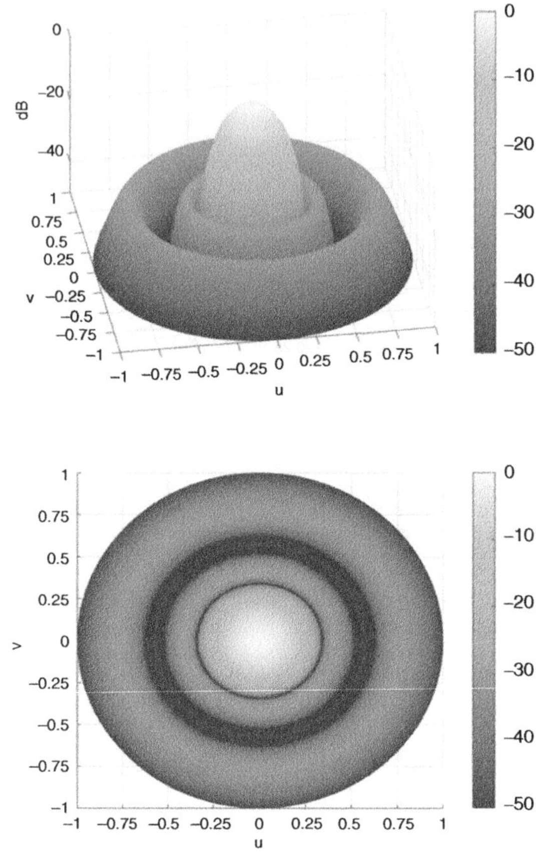

Figure 2.52 Similar to the rectangular grid, by applying amplitude weighting, the SLs can be reduced. For this example a Taylor weighting has been applied decreasing in magnitude from the first ring of elements to the last ring of elements in Figure 2.47.

transformations can be described using rotational matrices. The expressions in Equation 2.57 show the transformation matrices for roll, pitch, and yaw.

$$Roll = [R] = \begin{vmatrix} cos\theta_R & -sin\theta_R & 0 \\ sin\theta_R & cos\theta_R & 0 \\ 0 & 0 & 1 \end{vmatrix}$$

$$Pitch = [R] = \begin{vmatrix} 1 & 0 & 0 \\ 0 & cos\theta_P & sin\theta_P \\ 0 & -sin\theta_P & cos\theta_P \end{vmatrix} \tag{2.57}$$

$$Yaw = [Y] = \begin{vmatrix} cos\theta_Y & 0 & -sin\theta_Y \\ 0 & 1 & 0 \\ sin\theta_Y & 0 & cos\theta_Y \end{vmatrix}$$

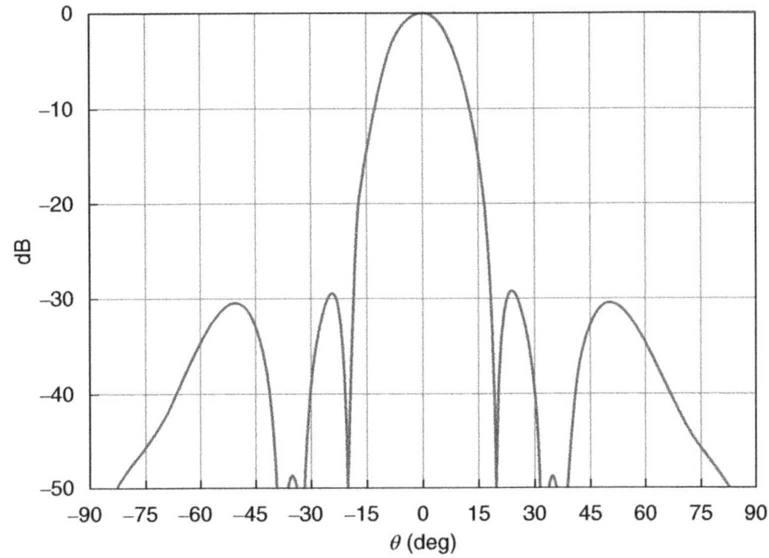

Figure 2.53 SL reduction for a weighted circular array. Similar to a rectangular grid, applying amplitude weighting decreases the SLLs.

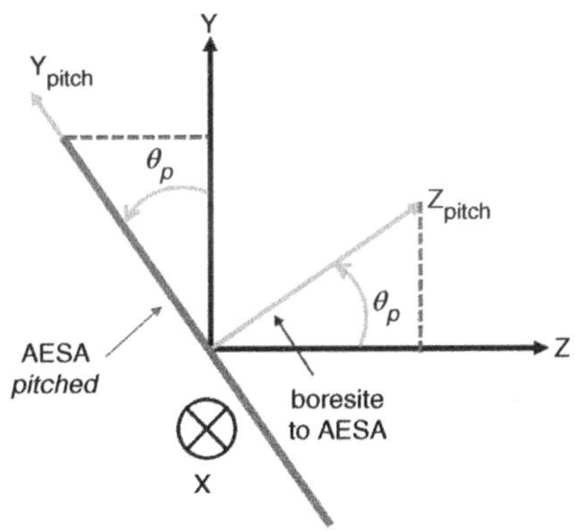

Figure 2.54 Example of an ESA tilted relative to the boresite of radar system coordinates.

These rotation matrices assume the roll, pitch and yaw angle directions are based on the right-hand rule. As an example, for a yaw rotation, the cross product of the unit vectors in the z- and x-axis generates a y-directed unit vector that is positive. This is illustrated in Figure 2.56.

Figure 2.57 shows a plot of a pattern without any pitch and a plot of a pattern with 30° pitch. As expected, the main beam of the pattern is shifted up 30° due to the pitch of the AESA. Corresponding plots for arbitrary roll, pitch, and yaw can be generated using the matrices defined in Equation 2.57. It is important to note that the order of the roll, pitch, and yaw matters. For example, $[R]\Delta[P] \neq [P]\Delta[R]$.

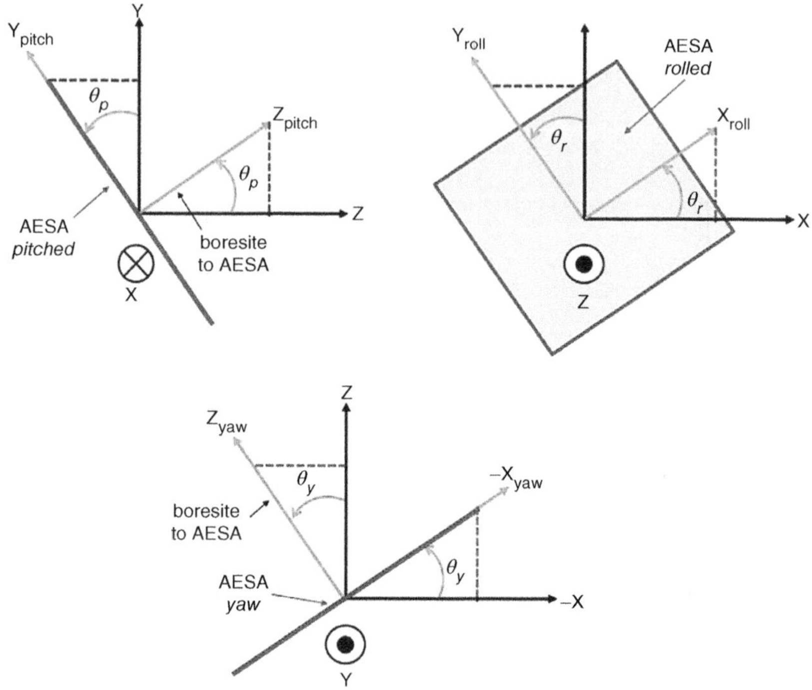

Figure 2.55 Roll, pitch, and yaw definitions.

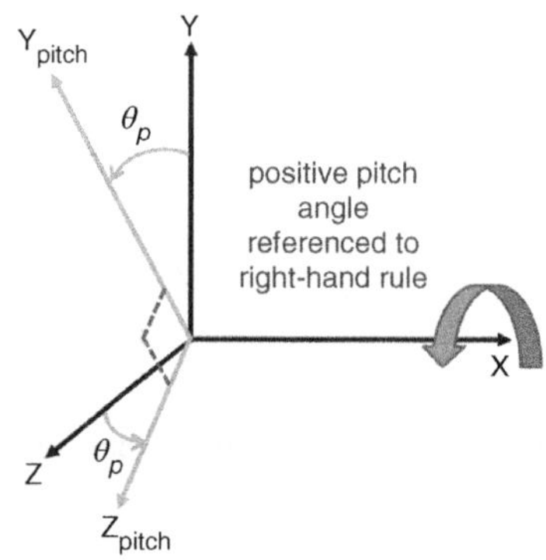

Figure 2.56 The right-hand rule dictates the direction of the roll, pitch, and yaw rotations.

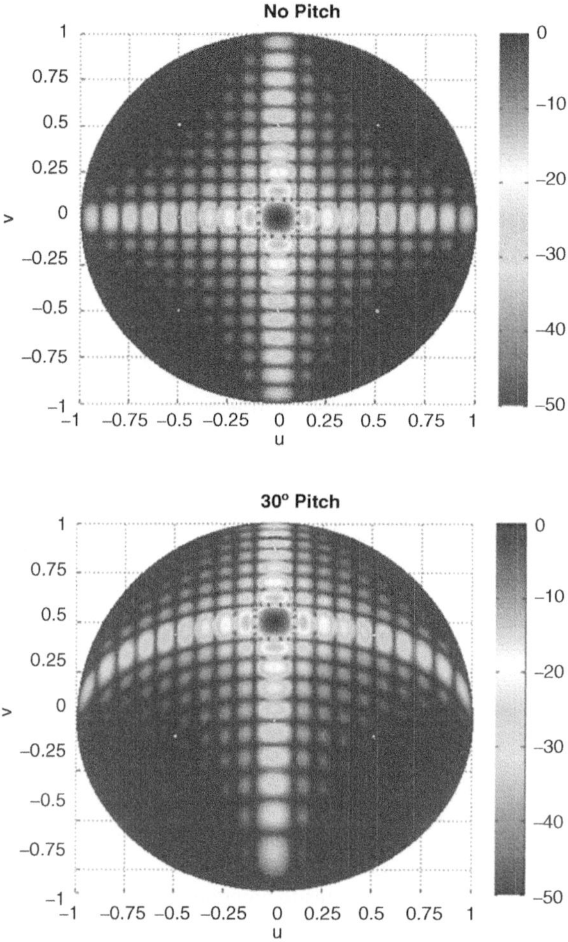

Figure 2.57 Effects of pitch on an AESA pattern.

2.9 Integrated Gain

The directive gain of an antenna is the ratio of the radiation intensity (power per unit solid angle) in a particular direction to the average power radiated over all space (Balanis, 1982). The equation for the directive gain is

$$D(\theta, \phi) = \frac{4\pi U(\theta, \phi)}{\int_{\phi=0}^{2\pi} \int_{\theta=0}^{\pi} U(\theta, \phi) sin\theta d\theta d\phi}. \tag{2.58}$$

The directivity of an antenna is the maximum directive gain value and can be written as

$$D = max(D(\theta, \phi)). \tag{2.59}$$

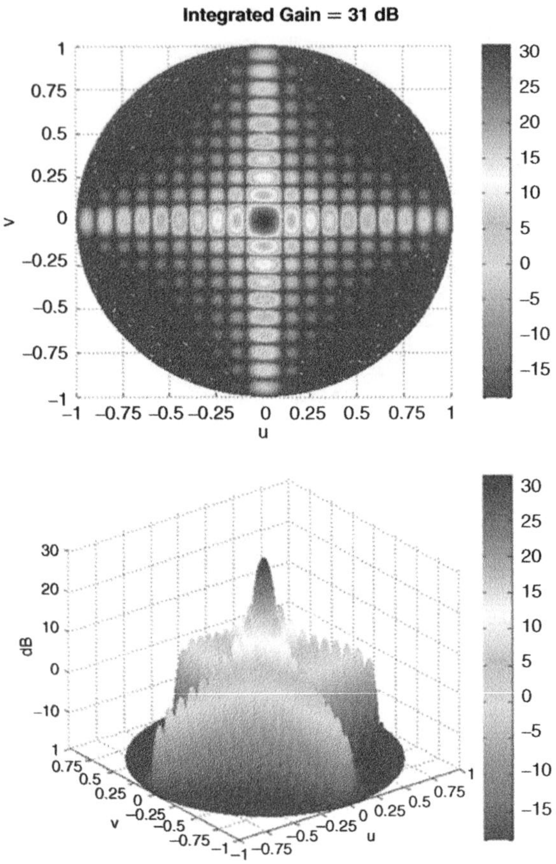

Figure 2.58 Integrated gain of the boresite antenna pattern (no electronic scan) in sine space shown in Figure 2.43.

An alternative and commonly used expression for the directive gain is

$$D = \frac{4\pi A}{\lambda^2} \cdot TL. \tag{2.60}$$

For large arrays, Equation 2.60 can be used. However, for small arrays Equation 2.59 provides a more accurate result. Figure 2.58 illustrates the integrated gain for the pattern in Figure 2.42. Using Equation 2.60, the peak directivity is 31 dB. The simulated integrated gain value is 31 dB.

References

Balanis, C. *Antenna Theory Analysis and Design*. John Wiley & Sons, Publishers, Inc., 1982.

Batzel, U., Ramsey, K., and Davis, D. "Angular coordinate definitions." Northrop Grumman Electronic Systems Antenna Fundamentals Course, 2010.

Davis, D. "Grating lobes analysis." Northrop Grumman Electronic Systems Antenna Fundamentals Course, 2010.

Holter, H., and Steyskal, H. "On the size requirement for finite phased-array models." *IEEE Transactions on Antennas and Propagation*, pp. 836–840, 2002.

Konapelsky, R. "Coordinate transformations." Northrop Grumman Electronic Systems Antenna Systems Class, 2002.

Long, J., and Schmidt, K. "Presenting antenna patterns with fft and dft." Northrop Grumman Electronic Systems Antenna Fundamentals Course, 2010.

Mailloux, R.J. *Phased Array Antenna Handbook*. Artech House, Norwood, MA, 1993.

Miller, C.J. "Minimizing the effects of phase quantization errors in an electronically scanned array." *Proc. 1964 Symp. Electronically Scanned Phased Arrays and Applications*, pp. 17–38, 1964.

Skolnik, M.I. *Radar Handbook*. McGraw Hill, 1990.

Taylor, T.T. "Design of line-source antennas for narrow beamwidth and low side lobes." *Transactions of the IRE Professional Group on Antennas and Propagation*, 3:16–28, 1955.

3

Array Elements

Key Concepts

- Bandwidth
- Polarization
- Array Grid
- Mismatch and Ohmic Loss
- Active Match
- Scan Loss

3.1 Introduction

The antenna array elements in an AESA are the conduit interface between the AESA and free space. On transmit, the array elements are the last RF component prior to free space transmission, and on receive they are the entry point for the energy from incoming signals from free space. Poorly designed array elements can affect: AESA gain when scanning, polarization performance for transmitting and receiving, and transmit and receive efficiency due to power loss. Because of this, key performance requirements are flowed to the antenna designer to ensure required system performance is achieved. Figure 3.1 highlights the array elements in the AESA.

In Chapter 1, the equation for SNR was derived. The primary effect of the array elements on the system SNR is the signal power component, which was shown to be:

$$S = \frac{P_{TX} G_{TX}^2 \sigma \lambda^2}{(4\pi)^3 R^4} (W). \tag{3.1}$$

Active Electronically Scanned Arrays: Fundamentals and Applications, First Edition. Arik D. Brown.

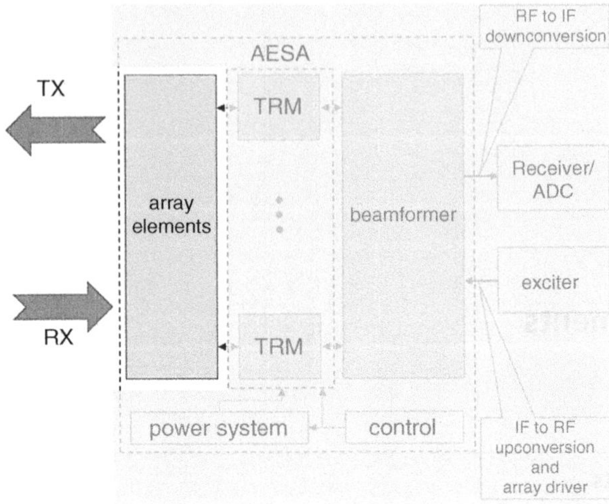

Figure 3.1 The antenna array elements serve as the interface between the AESA and free space/environment.

In Equation 3.1, the transmitted power, P_{TX}, and the transmit antenna gain, G_{TX}, are directly affected by the array element design. There are no amplifiers after the array elements, so any losses in the array element will result in a direct loss in signal power. This power loss can be quantified by mismatch loss and ohmic loss, which will be discussed in subsequent sections. Fundamentally, mismatch loss quantifies how much power is delivered to the array element from the TRM. If both have the same impedance at their interface, then all power would be transmitted and none would be reflected. However, in reality, both array element and TRM have different impedances, and there is reflected power, which is quantified by the mismatch loss. Ohmic loss is a function of the dielectric material contained in the element feed and array element in addition to the metal used to deliver power to the array element via stripline, microstrip, or waveguide input transmission.

Additionally, when designing array elements, consideration must be given to their operation in an array environment. What this means is that the mismatch loss for an isolated single element will be substantially different than the mismatch loss for an array element (Skolnik, 1990). This is depicted in Figure 3.2, which shows that for an array environment each element receives coupled energy from its neighboring elements in addition to its own reflected power. The mismatch loss in an array environment is referred to as active match. This will also be discussed more in depth in a later section.

In addition to the transmitted power (P_{TX}), the array elements are also a major driver for the antenna gain (G_{TX}) performance. The array elements must provide

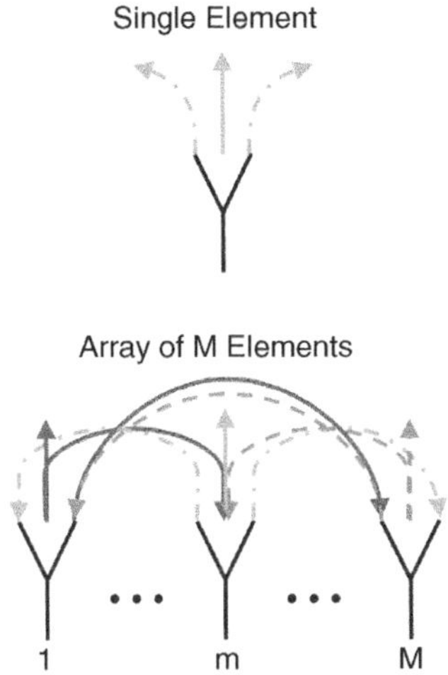

Figure 3.2 The single element has no additional coupled energy and its mismatch loss is unaffected. The array element has coupled energy from neighboring elements, which affects its mismatch loss and must be accounted for in the design.

the required antenna gain over the operational bandwidth, polarization, and AESA scan requirements. As AESAs have advanced, these requirements have become increasingly challenging. Many multi-function systems that require AESAs have one or more of the following requirements: multi-octave bandwidth, dual polarization, and required scan of 60° or greater. This can make array element design challenging.

For systems that are receive only (ESM, SIGINT), the array elements only affect the receive antenna gain since the power incident on the face of the AESA is from the incoming signal. Rewriting the SNR equation from Chapter 1, the antenna gain and system noise temperature can be isolated as a separate term. This is shown in Equation 3.2 where ERP is the product of the incoming signals antenna gain and transmit power.

$$SNR = ERP_{external} \cdot \left(\frac{\lambda}{4\pi R}\right)^2 \cdot \frac{1}{kBL} \cdot \frac{G}{T}. \tag{3.2}$$

This term, $\frac{G}{T}$, is typically referred to as *G over T*, where for many conditions $T = T_0F$. This is the primary parameter for receive sensitivity performance. So, no matter the application, the array element is critically important for providing optimal system performance.

It is important to take a moment to further examine Equation 3.2. A common mistake the author has seen made is assuming that by increasing the frequency, the sensitivity/SNR in Equation 3.2 increases as well. This *is not* the case. The determining factor in receive sensitivity is the equivalent aperture area A. This can be shown by rewriting Equation 3.2 with G replaced by the standard definition for gain ($G = \frac{4\pi A}{\lambda^2}$):

$$SNR = ERP \cdot \left(\frac{\lambda}{4\pi R}\right)^2 \cdot \frac{1}{kBL} \cdot \frac{\frac{4\pi A}{\lambda^2}}{T}, \tag{3.3}$$

which can be further simplified as:

$$SNR = \left(\frac{ERP}{4\pi R^2}\right) \cdot \frac{1}{kBL} \cdot \frac{A}{T}. \tag{3.4}$$

Because A in Equation 3.4 is the *effective* area, and not the physical area, $\frac{G}{T}$ is a better performance parameter since it can be readily measured in an antenna range for verification.

3.2 Bandwidth

Bandwidth is an important parameter that drives the antenna array element design. It is important to discuss the types of bandwidth before quantifying array element performance. The primary definitions for bandwidth relative to an AESA are operational bandwidth and instantaneous bandwidth (IBW). Both of these define the frequency boundaries that the AESA must perform over as a function of frequency.

Operational bandwidth describes the frequency range for which the AESA and the overall system must operate. In practice both a minimum and maximum frequency are flowed to the array element designer, which are derived from the operational bandwidth. As an example, consider an AESA radar system that must operate from 8 GHz to 9 GHz. The operational bandwidth here is 1 GHz. What this means is that the radar will transmit and receive energy over any subset of the operational bandwidth of 1 GHz.

The IBW is a subset of the operational bandwidth and defines for an instance of time the fractional bandwidth over which the system must perform within the operational bandwidth. Using the AESA radar example above, the system IBW will be selected to be 100 MHz, as an example. This means that the system will need to transmit and receive over 100 MHz anywhere within the operational bandwidth of 8 GHz to 9 GHz. Figure 3.3 provides a graphical illustration of operational bandwidth and IBW. As will be shown in Chapters 4 and 5, the IBW is a driving requirement for the transmit receive modules and the beamformer. However, it is not a requirement for the array elements. For the array elements, the primary requirements driver is the operational bandwidth.

It is critical that the array elements provide the required gain over frequency for the overall system to operate. As an example, a radar that employs frequency hopping or changes frequency to improve detection cannot sacrifice gain over the operational bandwidth. This is depicted in Figure 3.4.

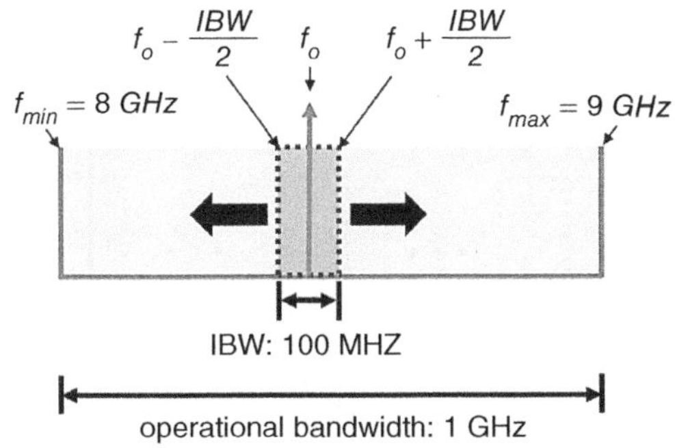

Figure 3.3 An example illustrating operational bandwidth and IBW. In the figure a 1 GHz operational bandwidth is specified from 8 GHz to 9 GHz. The IBW of 100 MHz defines what portion of the operational spectrum the system must function over within the operational bandwidth.

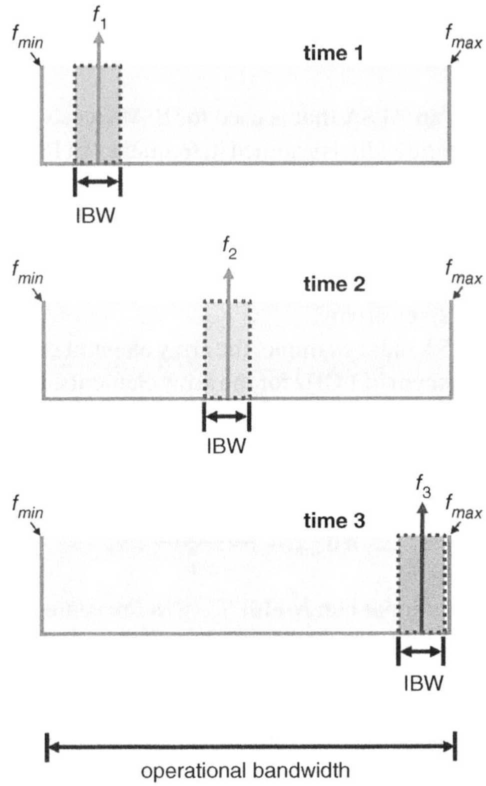

Figure 3.4 Illustration of a radar that changes frequency over the IBW. The array elements must provide the required gain over the entire operational frequency band to not degrade system performance.

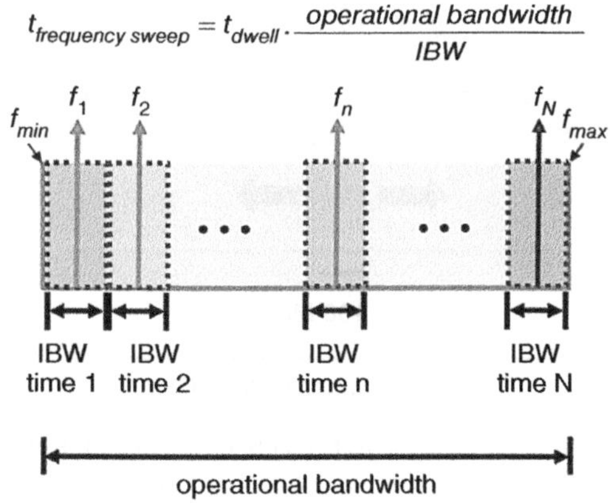

Figure 3.5 Illustration of an ESM system where the operational frequency band is scanned in IBW-sized swaths. Similar to the example in Figure 3.4, the array elements must provide the required gain over the entire operational frequency band to not degrade system performance.

Another example is an AESA that is used for ESM (receive only). Typically, for ESM the operational bandwidth is scanned in frequency in IBW-sized chunks with each IBW chunk having an associated dwell time. As an example, if the operational bandwidth is 1GHz and the IBW is 100 MHz, the system will sweep over the operational bandwidth with ten IBW swaths. This is shown in Figure 3.5. The array element gain must be sufficent over the entire operational bandwidth to ensure no degradation in SNR/sensitivity.

Referencing the AESA radar example, the array element designer would receive a flowed down requirement of 1 GHz for the array elements. In evaluating whether this requirement can be met with a simple array element design or something more complex, a fractional bandwidth definition is applied to the operational bandwidth and is shown in Equation 3.5,

$$BW_{frac} = \frac{f_{max} - f_{min}}{f_{center}}, \tag{3.5}$$

where BW_{frac} is the fractional bandwidth, f_{center} is the center frequency $(\frac{f_{min}+f_{max}}{2})$, f_{max} is the maximum frequency, and f_{min} is the minimum frequency. In practice, fractional bandwidths exceeding 20% are considered wideband and fractional bandwidths exceeding 50% are considered ultrawideband. Wideband and ultrawideband array element designs are challenging because the element's impedance match must be sufficient over frequency and AESA scan angle. This will be discussed further in Section 3.6.

3.3 Polarization

Polarization is a characterization of electromagnetic field propagation as a function of time. It is important for AESAs because it describes the polarization of the transmitted/received energy from/by the array elements. Because of the Lorentz reciprocity theorem (Balanis, 1982), the polarization response of an antenna is the same whether transmitting or receiving; thus, polarization here will be discussed in a general sense and not specific to either the transmit or receive operation.

There are three different types of polarization: random, partial, and complete (Stutzman, 1993). Randomly polarized electromagnetic waves change their polarization randomly as a function of time. Partially polarized waves vary between deterministic and randomly changing as a function of time, and completely polarized waves change deterministically as a function of time. Antennas are considered completely polarized, and this will be the focus of this section.

At the system level, polarization ultimately can be related to loss or SNR degradation. As an example in communications, if the transmit antenna and the receive antenna do not have the same polarization, then the received energy will be degraded. This can be shown by modifying the SNR equation for a one-way transmit to receive antenna arrangement as shown in Equation 3.6:

$$SNR = ERP \cdot \left(\frac{\lambda}{4\pi R}\right)^2 \cdot \frac{1}{kBL} \cdot \frac{G}{T} \cdot (\hat{r} \cdot \hat{p}), \tag{3.6}$$

where $\hat{r}$ is the unit vector of the transmitted polarization, and $\hat{p}$ is the polarization of the receive polarization. Since $\hat{r} \cdot \hat{p} = cos(\theta_{rp})$, where θ_{rp} is the angle between the polarization vectors, if the polarizations are orthogonal ($\theta_{rp} = 90°$), or simply not aligned ($\theta_{rp} \neq 0°$), the SNR will be degraded.

For radar, polarization can also help improve the detection of signals for enhanced target discrimination (Skolnik, 2001) or to detect and characterize weather clutter via radar polarimetry. By replacing the ERP term in Equation 3.6 with the expression $ERP \cdot \frac{\sigma}{4\pi R^2}$, the same conclusions can be made for the radar case.

Before describing how polarization can be characterized for an AESA, it is first important to cover some basic electromagnetic fundamentals and how polarization can be represented for analysis purposes. This will naturally lead to how polarization affects the AESA and more specifically the array elements. Depending on the system polarization requirements, polarization can have a huge effect on the complexity of the array element design. The following three sections will cover electromagnetic polarization fundamentals, types of polarization, expressions for representing polarization states, and array polarization. For a more in-depth examination of polarization, (Stutzman, 1993) is recommended.

3.3.1 Electromagnetic Polarization Fundamentals

The power that is described in the RRE is due to electromagnetic waves propagating spatially as a function of time. As electric and magnetic fields propagate at distances far from their source, they travel as plane waves. This means the fields are a function of the vector components that are perpendicular to the direction of propagation. The distance at which the fields can be categorized as plane waves is the far-field distance. This distance can be calculated as $\frac{2D^2}{\lambda}$, where D is the longest dimension of the antenna source, and λ is the wavelength. The polarization discussion in this book will refer to plane wave propagation.

The time-varying expressions for the electric and magnetic fields can be represented as

$$\vec{E}(t,z)\ \left[\frac{V}{m}\right] \tag{3.7}$$

and

$$\vec{H}(t,z) = \frac{1}{\eta_o}\hat{n} \times \vec{E}(t,z)\ \left[\frac{A}{m}\right], \tag{3.8}$$

where $\vec{E}(t,z)$ is the electric field, $\vec{H}(t,z)$ is the magnetic field, η_o is the free space impedance ($\eta_o = \sqrt{\frac{\mu_o}{\epsilon_o}} = 377\ \Omega$), and the direction of propagation is assumed to be z (Stutzman, 1993). With the expressions for the electric and magnetic fields in Equations 3.7 and 3.8, the power associated with the fields can be represented by the Poynting vector (Balanis, 1982)

$$\vec{S}(t,z) = \vec{E}(t,z) \times \vec{H}(t,z)\ \left[\frac{W}{m^2}\right] \tag{3.9}$$

that represents the direction and magnitude of the propagated power. Substituting Equation 3.8 into Equation 3.9, the electromagnetic power can be represented solely as a function of the electric field:

$$\vec{S}(t,z) = \vec{E}(t,z) \times \left(\frac{1}{\eta_o}\ \hat{n} \times \vec{E}(t,z)\ \left[\frac{W}{m^2}\right]\right. . \tag{3.10}$$

Assuming the direction of propagation is $\hat{z}$ and that the electric and magnetic fields have only x and y vector components as plane waves, Equation 3.10 can be rewritten as:

$$\vec{S}(t,z) = \hat{z}\frac{|\vec{E}(t,z)|^2}{\eta_o}\ \left[\frac{W}{m^2}\right]. \tag{3.11}$$

This is important because polarization analysis for antennas uses the electric field variation. With only the calculated electric field, the polarization can be characterized completely. Another important observation is that the expression $\frac{ERP}{4\pi R^2}$ used in the RRE is directly attributable to the Poynting vector, which has units of $\frac{W}{m^2}$.

3.3.2 Types of Polarization

Polarization of a radiated wave is defined as "that property of a radiated electromagnetic wave describing the time varying direction and relative magnitude of the electric-field vector; specifically, the figure traced as a function of time by the extremity of the vector at a fixed location in space, and the sense in which it is traced, as observed along the direction of propagation" (Balanis, 1982). In more simple terms, it is the motion of the tip of the instantaneous electric field vector with time at a fixed point in space (Stutzman, 1993). Polarization can be classified into three different categories: linear, circular, and elliptical. For linear polarization, the electric field traces a straight line as a function of time; hence the name linear. For circular polarization the electric field traces out a circle as a function of time and similarly, an ellipse for elliptical polarization. Circular polarization is a form of elliptical polarization where both of the electric field components are equal. For both circular and elliptical polarization there are two senses of rotation, CW or CCW, which describes the direction of the rotation as a function of time.

To better understand the polarization types, we start with the following expression for the time-varying electric field:

$$\vec{E}(t,z) = \vec{E}_x\hat{x} + \vec{E}_y\hat{y}\ \left[\frac{V}{m}\right]. \tag{3.12}$$

In Equation 3.12, the vector components of the electric field are complex. The instantaneous components of the electric field vector can be related to the complex components using the following expressions (Balanis, 1982):

$$\vec{E}_x = E_x cos(\omega t + kz + \phi_x)\hat{x}\ \left[\frac{V}{m}\right] \tag{3.13}$$

and

$$\vec{E}_y = E_y cos(\omega t + kz + \phi_y)\hat{y}\ \left[\frac{V}{m}\right]. \tag{3.14}$$

E_x and E_y are the maximum magnitudes of the electric field components, ω is the frequency, k is the propagation constant ($\frac{2\pi}{\lambda}$), and ϕ_x and ϕ_y are the respective phases of the vector components. Slant linear polarization is depicted in Figure 3.6 as a function of variable distance z and fixed time t.

3.3.2.1 Linear Polarization

For linear polarization, the phase of the electric field components are equal ($\phi_x = \phi_y$). For ease of notion both ϕ_x and ϕ_y can be set to zero, so Equations 3.13 and 3.14 become:

$$\vec{E}_x = E_x cos(\omega t + kz)\hat{x}\ \left[\frac{V}{m}\right] \tag{3.15}$$

and

$$\vec{E}_y = E_y cos(\omega t + kz)\hat{y}\ \left[\frac{V}{m}\right]. \tag{3.16}$$

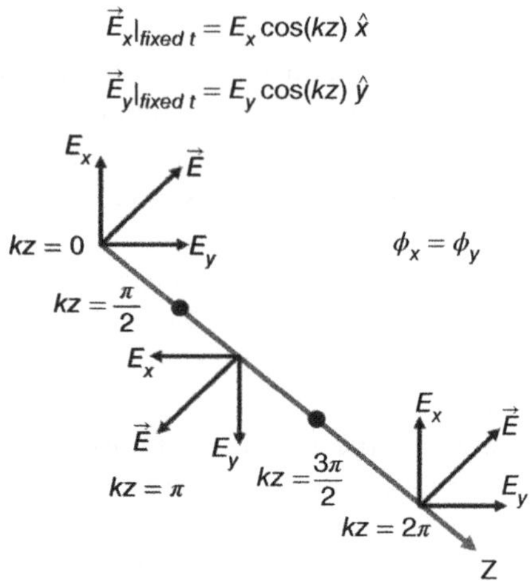

Figure 3.6 The figure highlights how the electric field orientation changes as a function of propagation distance z for a fixed instance of time. The figure is an example of slant linear polarization where $\phi_x = \phi_y$.

In general, linear polarization results when the delta phase between the electric field components ($\Delta\phi$) can be described with the following expression (Balanis, 1982):

$$\Delta\phi = \phi_y - \phi_x = \pm n\pi, \quad n = 0, 1, 2, \ldots \tag{3.17}$$

When n is equal to 0 or multiples of 2π, the electric field traces a line with a positive slope, and when n is an odd multiple of π, the electric field traces a line with a negative slope, as depicted in Figure 3.7.

The ratio of the electric field component magnitudes describes the orientation of the line that the electric field traces and is called the tilt angle (τ). The tilt angle for linear polarization can be expressed as

$$\tau = tan^{-1}\left(\frac{E_y}{E_x}\right). \tag{3.18}$$

For τ equal to 0° or π, it is horizontal linear polarization; for τ equal to $\pm\frac{\pi}{2}$, it is vertical polarization; and for $\tau \neq 0, \pi$, or $\frac{\pi}{2}$, it is slant linear polarization.

3.3.2.2 Circular Polarization

Circular polarization occurs when $\Delta\phi$ equals $\pm\frac{\pi}{2}$, and when E_x and E_y are equal. In this case, the electrical field traces out a circle for any fixed position along the direction of propagation. When $\Delta\phi$ is equal to $\frac{\pi}{2}$, this is referred to as right-hand circular polarization (RHCP). The y component of the electric field leads the x component by $\frac{\pi}{2}$. When $\Delta\phi$ is equal to $-\frac{\pi}{2}$, this is referred to as left-hand circular

$$\vec{E}_x|_{fixed\ z} = E_x \cos(wt)\ \hat{x}$$

$$\vec{E}_y|_{fixed\ z} = E_y \cos(wt)\ \hat{y}$$

$$\phi_x = \phi_y$$

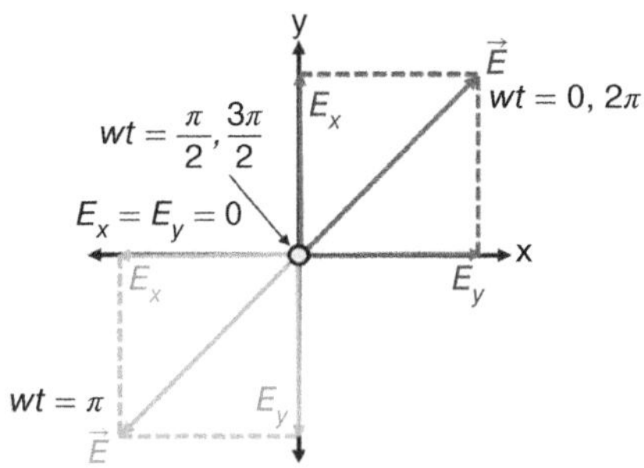

Figure 3.7 The polarization vector for a fixed spatial point traces out a line as a function of time for linear polarization.

polarization (LHCP). More generally, the phase relationship between electric field components can be expressed as

$$\Delta\phi = \phi_y - \phi_x = \begin{cases} +\left(\frac{1}{2} + 2n\right)\pi, & n = 0, 1, 2, \ldots\ for\ RHCP \\ -\left(\frac{1}{2} + 2n\right)\pi, & n = 0, 1, 2, \ldots\ for\ LHCP \end{cases}. \tag{3.19}$$

The orientation for RHCP is clockwise, using the right-hand rule in the direction of propagation, while LHCP is counterclockwise in the direction of propagation.

3.3.2.3 Elliptical Polarization

For an AESA array element design, elliptical polarization is not a design driver. Array elements for AESAs are typically designed to support horizontal, vertical, slant linear, RHCP, LHCP, and dual polarization (a combination of both horizontal and vertical polarization or RHCP and LHCP). The reason for this is that any polarization can be represented with two orthogonal polarizations. However, it is important to understand elliptical polarization because as the array element is scanned, its polarization changes, and this results in elliptical polarization. This can be important for an AESA that has strict polarization requirements over all scan angles and must be calibrated.

Elliptical polarization occurs in the following two cases. First, when the electric field components are not equal $(E_x \neq E_y)$ *and* $\Delta\phi \neq 0$. Second, elliptical polarization occurs when $E_x = E_y$ *and* $\Delta\phi \neq \frac{\pi}{2}$. More generally, these two cases can be expressed as (Balanis, 1982)

$$E_x \neq E_y, \Delta\phi = \phi_y - \phi_x = \begin{cases} +\left(\frac{1}{2} + 2n\right)\pi, & n = 0, 1, 2, \ldots \textit{ for RHEP} \\ -\left(\frac{1}{2} + 2n\right)\pi, & n = 0, 1, 2, \ldots \textit{ for LHEP} \end{cases} \tag{3.20}$$

or

$$E_x = E_y, \Delta\phi = \phi_y - \phi_x \neq \pm\frac{n}{2}\pi \begin{cases} > 0, & n = 0, 1, 2, \ldots \textit{ for RHEP} \\ < 0, & n = 0, 1, 2, \ldots \textit{ for LHEP} \end{cases} \tag{3.21}$$

where RHEP and LHEP represent right-hand elliptical polarization and left-hand elliptical polarization, respectively. Figure 3.8 shows a summary view of slant linear, RHEP, and LHEP.

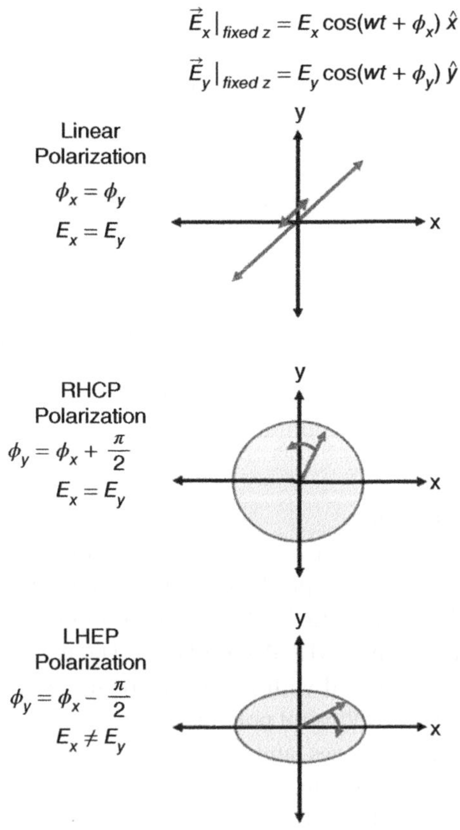

Figure 3.8 Graphical expressions for the different types of polarization. Circular polarization is a subset of elliptical polarization and looks similar except that the electric field components are equal in magnitude.

3.3.3 Polarization States

Polarization state representations have been succinctly defined in (Stutzman, 1993). The polarization state representations are:

1. Polarization Ellipse (ϵ, τ)
2. Polarization Ellipse (γ, $\Delta\phi$)
3. Poincare Sphere
4. Complex Vector
5. Stokes Parameters
6. Polarization Ratio

All of these representation are similar in that they uniquely define the polarization of the electric field. For this chapter, only the polarization ellipse representations (this section) and the complex vector representation (next section) will be discussed. The reader is encouraged to refer to (Stutzman, 1993) for an in-depth discussion on all the states. Because of the simplicity for how they can be constructed, we will focus on the polarization ellipse representation. The complex vector representation is typically used to evaluate the simulated and/or measured response of the electric field for polarization performance. This will be described in Section 3.3.4.

For the first polarization ellipse representation there are two angles (ϵ and τ), which must be defined. Both of these angles are shown in Figure 3.9. The tilt angle τ describes the orientation of the polarization ellipse around the x- and y-axis and can be defined as (Stutzman, 1993):

$$0 \leq \tau \leq 180^\circ. \tag{3.22}$$

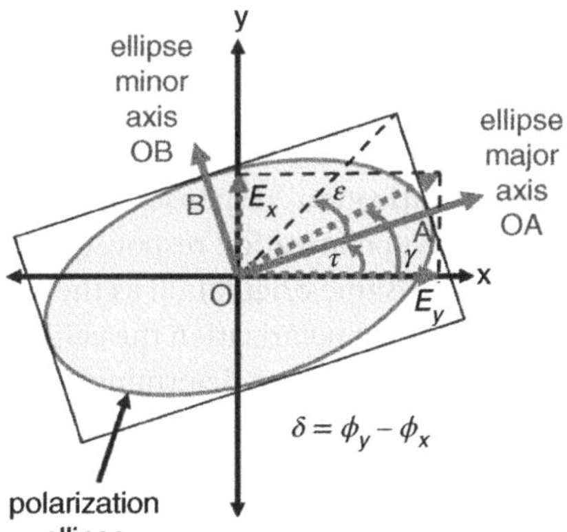

Figure 3.9 Generalized polarization ellipse expressing polarization for a fixed stationary point along the direction of propagation (Stutzman, 1993).

To define the ellipticity angle ϵ, we first need to define the axial ratio (AR). The axial ratio is a measure of the ellipse shape and can be represented as (Stutzman, 1993):

$$|AR| = \frac{major\ axis\ length}{minor\ axis\ length} = \frac{OA}{OB} \geq 1. \tag{3.23}$$

Once the AR is defined, ϵ can then be defined as (Stutzman, 1993):

$$\epsilon = cot^{-1}(-AR) \quad -45^\circ \leq \epsilon \leq 45^\circ, \tag{3.24}$$

where right-hand sense is $sign(AR) = +$ and for left-hand sense $sign(AR) = -$. The angle pair (τ, ϵ) completely represents all polarization states.

The second polarization ellipse representation completely and uniquely describes the polarization state with two different angles. The first angle has already been discussed and is $\Delta\phi$, the phase delta between the polarization of the electric field vector components. The second angle, γ, shown in Figure 3.9, represents the relationship between the electric field amplitude components and is defined by (Stutzman, 1993):

$$\gamma = tan^{-1}\left(\frac{E_y}{E_x}\right). \tag{3.25}$$

This angle pair $(\Delta\phi, \gamma)$ completely defines the polarization state.

Using the following relationships, both (τ, ϵ) and $(\Delta\phi, \gamma)$ can be derived from one another.

$$\begin{aligned} sin(2\epsilon) &= sin(2\gamma)sin(\Delta\phi) \\ tan(2\tau) &= tan(2\gamma)cos(\Delta\phi) \end{aligned} \tag{3.26}$$

Figure 3.10 shows a the graphical representation of a subset of the range of polarization ellipse shapes that are represented using (τ, ϵ) and $(\Delta\phi, \gamma)$. The full set can be found in (Stutzman, 1993).

3.3.4 Array Polarization

When verifying the performance of an array of elements in an AESA, the co-polarization and cross-polarization responses of the array are measured, analyzed, and evaluated. The co-polarization response is the response of another reference antenna aligned in the same orientation as the array. As an example, for an array that is designed for vertical polarization the calibrated reference antenna used would also be vertically polarized to determine how pure the polarization of the array is. In practice, this is never perfect, and there is always energy in the orthogonal or cross-polarization. This is measured by rotating the reference antenna so that its polarization is orthogonal to the polarization of the antenna under test (AUT). The cross-polarization response is then measured. Once both the co-polarization and cross-polarization responses are known, the polarization performance of the array is known.

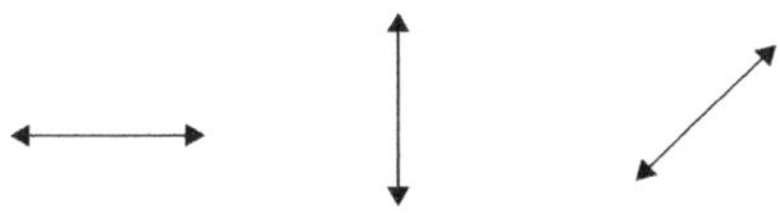

Pol. State	Horizontal	Vertical	Slant Linear
ε, τ	0°, 0°	0°, 90°	0°, 45°
δ, γ	0°, 0°	0°, 90°	0°, 45°

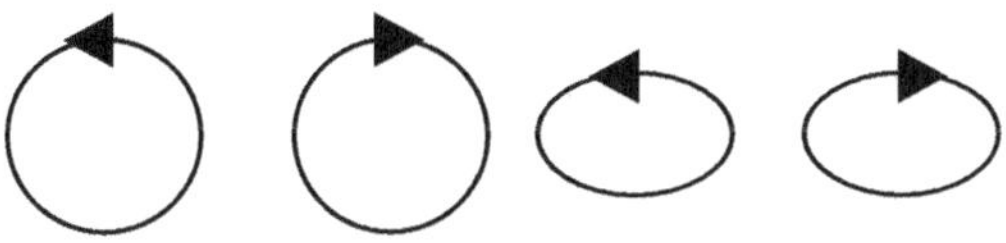

Pol. State	RHCP	LHCP	RHEP	LHEP
ε, τ	–45°, 0°	+45°, 0°	–22.5°, 0°	+22.5°, 0°
δ, γ	–90°, 45°	+90°, +45°	–90°, 22.5°	+90°, 22.5°

Figure 3.10 Visual aids provide insight into the polarization state representations. (Stutzman, 1993).

Although measuring the co- and cross-polarization response of the array is frequently performed in an antenna range, this can also be done with a high-fidelilty numerical analysis simulator tool. The array element is modeled with array boundary conditions to model the element pattern response as a function of frequency and scan angle. The magnitude and phase of the element pattern electric field can then be used to evaluate the polarization performance (i.e., if the array is designed for a certain polarization, how pure is that polarization versus frequency, angle, and scan angle).

To analytically model this performance, the complex vector representation for the measured or modeled electric field array element response is used. The following expressions are used to analyze the co- and cross-polarization performance of the array (Stutzman, 1993),

$$F_{co}(\theta, \phi) = \vec{E}(\theta, \phi) \cdot \hat{e}_{co} \tag{3.27}$$

$$F_{cross}(\theta, \phi) = \vec{E}(\theta, \phi) \cdot \hat{e}_{cross}, \tag{3.28}$$

where $\vec{E}$ is the measured or simulated complex electric field response, and $\hat{e}_{co}$ and $\hat{e}_{cross}$ are co- and cross-pol complex unit vectors for a given polarization state. In (Ludwig, 1973), polarization unit vector expressions were derived to evaluate polarization of an antenna. There are three different conventions used in (Ludwig, 1973), and they are commonly referred to as Ludwig 1, 2, and 3. When measuring antenna patterns, the Ludwig 3 expression is commonly used as it reflects how antennas are typically measured in an antenna range. As an example, the Ludwig 3 unit vectors for horizontal and vertical polarization are

Vertical Polarization

$$\begin{aligned}\hat{e}_{co} &= sin(\phi)\hat{\theta} + cos(\phi)\hat{\phi}\\ \hat{e}_{cross} &= cos(\phi)\hat{\theta} - sin(\phi)\hat{\phi}\end{aligned} \tag{3.29}$$

Horizontal Polarization

$$\begin{aligned}\hat{e}_{co} &= cos(\phi)\hat{\theta} - sin(\phi)\hat{\phi}\\ \hat{e}_{cross} &= sin(\phi)\hat{\theta} + cos(\phi)\hat{\phi}.\end{aligned} \tag{3.30}$$

An important factor to note is that in Equations 3.29 and 3.30, $\hat{e}_{co} \cdot \hat{e}_{cross} = 0$. So, when evaluating the polarization response of the array, if F_{cross} has a significant magnitude that is not much smaller than F_{co}, then the array's polarization response is poor.

3.3.4.1 Key Requirements

Based on the material covered in this section, there are several key polarization requirements that are flowed down to the array element designer. They are:

1. Polarization specification: This is simply the polarization that the array element needs to achieve. As an example, if the system requires vertical polarization, then a vertically polarized antenna design would be selected. If multiple polarizations are required such as dual horizontal and vertical, then the array element has to support both. This becomes a significant driver in the AESA complexity. For every element, the number of channels is doubled, which can be a challenge mechanically and also electrically for isolation.
2. Cross-polarization or orthogonality: This can be calculated by taking the magnitude of the ratio $\frac{F_{cross}}{F_{co}}$ and calculating the phase delta between F_{co} and F_{cross}. A ratio of –20 dB is normally achievable along the principal axes; however, for spatial points in the intercardinal regions it can be difficult to maintain magnitude and phase over scan.

Polarization isolation is also an important requirement. The electronics behind the array elements can degrade the cross-polarization if they allow significant coupling within the AESA. For systems that have two polarizations, it is essential that they remain isolated as they traverse through the AESA so that the received or transmitted polarization is not affected and orthogonality is maintained.

3.4 Array Grid

In Chapter 2, it was shown that periodic replicas of the array pattern called grating lobes are produced when the array element spacing is too large. The expression for the maximum element spacing was shown to be:

$$d = \frac{\lambda}{1 + sin\theta_o}. \tag{3.31}$$

As simple as this equation appears, it is one of the largest drivers in the cost of an AESA. For a typical AESA, there are high-power amplifiers (HPAs), phase shifters, attenuators, low-noise amplifiers (LNAs), circulators (for radars), and switches for each element. The electronic density increases dramatically since it is directly proportional to the number of array elements. In an AESA, the TRMs are typically the largest cost driver, and every effort is made to ensure that the AESA design uses the fewest number of elements possible to meet the requirements.

As an example, consider an x band AESA that operates in the range 8–10 GHz, and has a requirement to scan to 60°. As was discussed in Chapter 2, the highest frequency in the operational bandwidth is chosen to select the element spacing. Equation 3.31 can be rewritten to express the required element spacing as

$$d_{min} = \frac{c}{f_{max}(1 + sin\theta_o)}. \tag{3.32}$$

This shows that the smallest element spacing is required at the max operating frequency. Consider if instead of using Equation 3.31, the element spacing, d, was chosen conservatively to be $\frac{\lambda}{2}$. For a two-dimensional rectangular array with the same element spacing in both dimensions, the conservative element area would be

$$A_{elem_{conservative}} = d^2_{conservative} = \left(\frac{\lambda}{2}\right)^2, \tag{3.33}$$

while the element area based upon the element spacing in Equation 3.31 would be

$$A_{elem} = d^2 = \left(\frac{\lambda}{1 + sin(60°)}\right)^2. \tag{3.34}$$

The increase in the number of elements using the conservative approach can be calculated by taking the ratio of the element areas in Equations 3.33 and 3.34 expressing this as $\frac{4}{(1+sin(60°))^2}$. This represents a 15% increase in the number of elements, which translates into 15% more TRMs required, which would in turn be a substantial cost (see Figure 3.11).

The array grid also affects the impedance match of the antenna at large scan angles. As the main beam of the array is scanned close to its scan limits, the grating lobes begin to approach real space. This affects the impedance seen by the element and can cause the match to degrade, thereby reducing the efficiency and increasing the loss in the system. In order to combat this, the array element designer will

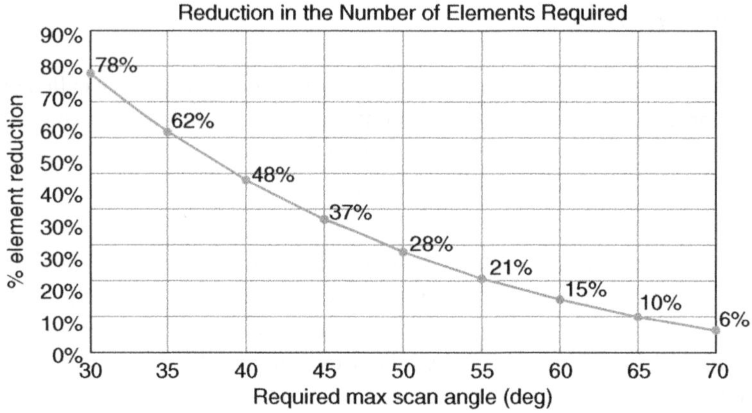

Figure 3.11 The graph shows the reduction in the number of elements required by using Equation 3.31 instead of $d = \frac{\lambda}{2}$. In many cases using $d = \frac{\lambda}{2}$ adds unnecessary conservatism and an overpriced, over-designed AESA. Because of this, $d = \frac{\lambda}{2}$ element spacing is used only for large scan angles (i.e., 70°) or in mmW AESAs where the cost per element is reduced due to wafer processing.

typically add margin to the required element spacing that translates to placing the grating lobe further away from the real space boundary to ensure the element match is not perturbed.

3.5 Mismatch and Ohmic Loss

Chapter 1 discussed how an AESA directly affects the RRE. P_{TX}, G (TX and RX), and T are all parameters in the RRE that are affected by an AESA. The array elements directly and indirectly affect all of these parameters as well. To this point, we have discussed operational bandwidth, polarization, and array grid, which are major contributors to AESA performance. However, they are not direct requirement flow downs from P_{TX}, G (TX and RX), and T but have their origin in the higher-level system requirements. This section and the next two address performance parameters that are direct flow downs from P_{TX}, G (TX and RX), and T.

In this section, mismatch loss and ohmic loss (also referred to as insertion loss) will be discussed. Both of these losses affect P_{TX}, G (TX and RX), and T. Figure 3.12, shows how mismatch and ohmic loss affect the power transmitted by the array element. Although the transmit operation is shown for illustration, the same explanation can be applied to receive operation as well.

Figure 3.12, represents a single-element channel in an AESA. For each element there is a TRM that sends amplified signal power from the HPA to the array

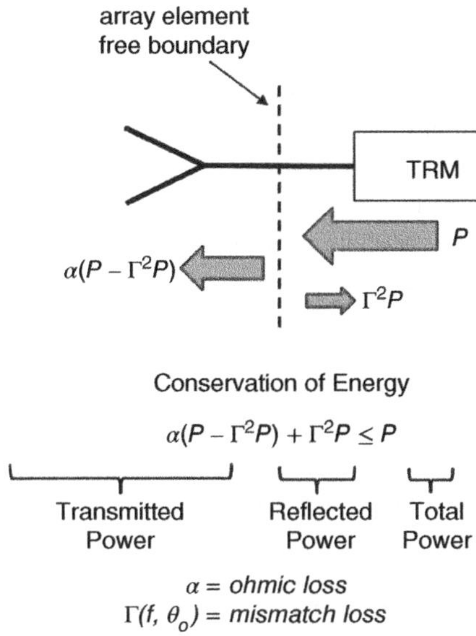

Figure 3.12 The diagram shows a transmit example where signal power is being transmitted from the TRM to the array element. At the TRM/array element boundary there is typically a connector that is matched to the impedance of the output of the TRM. Power is reflected from the boundary (reflection loss), and power is transmitted to the array element. Ohmic losses in the design of the array element can then degrade the transmitted power further. This same example applies on receive as well.

element. This power is denoted as P. The array element in many cases has some type of feed structure that is integrated in the array element that helps to create the polarization required for the AESA and also transforms the impedance from the TRM to the array element. This structure is typically integrated with the TRM via a connector or in some cases a microwave transition such as microstrip or stripline. At this boundary, the impedance from the TRM must be matched to that of the array element to prevent inefficient transmission of power.

In reality, there is always some power that is reflected back from this boundary and that is represented by a reflection coefficient Γ (Pozar, 2012). Γ represents the reflected voltage from the interface that has differing impedances (i.e., the TRM and the array element). In order to quantify the reflected power, Γ is squared resulting in $|\Gamma|^2$. The mismatch loss is then expressed as

$$\textit{mismatch loss} = |\Gamma|^2 P \tag{3.35}$$

and is depicted in Figure 3.12. In practice the TRM is given an output impedance requirement, and the array element must meet a mismatch loss requirement based on the TRM output impedance requirement. As will be discussed in the next section, Γ is not only a function of frequency but also of scan angle. Mismatch loss is important because it degrades the transmitted (and received) power of the array element. At the AESA system level this loss is typically allocated to array element losses in the signal chain. If the mismatch performance is poor, there will be a direct degradation of transmitted ERP and the gain on receive.

After power is transferred to the array element, there is another loss that has to be accounted for in the array element design. The power that is delivered to the array element is (Pozar, 2012):

$$transmission = (1 - |\Gamma|^2)P \tag{3.36}$$

and is shown in Figure 3.12. Due to conservation of energy the transmission power is simply the power delivered by the TRM, P, minus the mismatch loss ensuring that

$$mismatch\ loss + transmission = P \tag{3.37}$$

which can be rearranged to be

$$transmission = P - mismatch\ loss \tag{3.38}$$

where Equation 3.38 is the same as Equation 3.36.

The transmission power must traverse through the array element before reaching free space and will encounter ohmic losses due to the conductivity and permittivity of the array element. These losses are sometimes referred to as i^2R losses since this relates to the power loss due to currents traveling via a conductor through a material medium. Using α as a representation of the ohmic loss and Equation 3.39, the power radiated to free space is

$$transmission = \alpha(1 - |\Gamma(f, \theta_o)|^2)P, \tag{3.39}$$

which incorporates the effects of impedance and ohmic losses on the transmitted power.

The array element designer is given a mismatch loss and an ohmic loss requirement that flows directly from the ERP and/or $\frac{G}{T}$ requirements. These losses do not have to be accounted for separately and can be incorporated in the realized gain of the array element. The gain of the element represented by $\frac{4\pi A_{elem}}{\lambda^2}$ can be modified as (Mailloux, 1993):

$$Realized\ Gain = \alpha(1 - |\Gamma(f, \theta_o)|^2)\frac{4\pi A_{elem}}{\lambda^2}, \tag{3.40}$$

where the mismatch and ohmic loss are included in the realized gain expression. In practice, it doesn't matter which approach is taken as long as the mismatch and ohmic losses are included and are not "double-booked" (conveying a false degradation of performance).

Thus far, the focus has been on the loss impacts to power and gain of the AESA but not the noise temperature T. System engineers typically use T to quantify the effects of noise; however, antenna designers use an equivalent expression, which is noise factor, F. Both represent the added noise to a system that degrades the overall SNR. Using the expression

$$F = \frac{T_o + T}{To}, \tag{3.41}$$

the noise factor F and T can be related to each other where T_o is 290 K (room temperature) (Pettai, 1984). When converted to decibels, F is called noise figure (NF) and can be expressed as:

$$NF = 10log_{10}(F) \tag{3.42}$$

To understand how the mismatch and ohmic losses affect F the expression for cascaded F must be considered. An AESA has multiple stages of components (amplifiers, attenuators, switches, etc.) that are cascaded together contributing to an overall or effective noise factor F_{eff}. The expression for F_{eff} is written as (Pettai, 1984):

$$F_{eff} = F_1 + \frac{F_2 - 1}{G_1} + \frac{F_3 - 1}{G_1 G_2} + ... + \frac{F_N - 1}{G_1 G_2 ... G_{N-1}} \tag{3.43}$$

where N is the number of component stages and G_i is the gain at stage i ($i = 1, 2, 3, \dots N$).

The importance of Equation 3.43, is that if the first gain stage G_1 is selected to be large in value, then since its in the denominator of ($N - 1$) of the terms in Equation 3.43,

$$F_{eff} \approx F_1. \tag{3.44}$$

To take advantage of this, the first amplifier in the receive chain (LNA) is designed to have a large gain value, G_1, that is typically greater than 20 dB, which is a factor of 100 ($10log_{10}(100) = 20\ dB$). The F_1 term in Equation 3.44 represents the front-end losses that precede the LNA, which include the mismatch and ohmic loss. What is readily apparent from Equations 3.44 and 3.43 is that the front-end losses have a direct 1:1 contribution to the overall noise factor (F_{eff}). Thus, if the mismatch and ohmic losses are excessively large, they will degrade the overall $\frac{G}{T}$, which is undesirable.

3.6 Active Match

The mismatch loss described in the previous section characterizes the impedance mismatch of the array element. In this section, the expression for the reflection coefficient, Γ, in an AESA will be derived and is called the active match. The term *active* is used to denote that Γ in an array environment such as an AESA is different than the impedance match of an isolated antenna element. In general, the

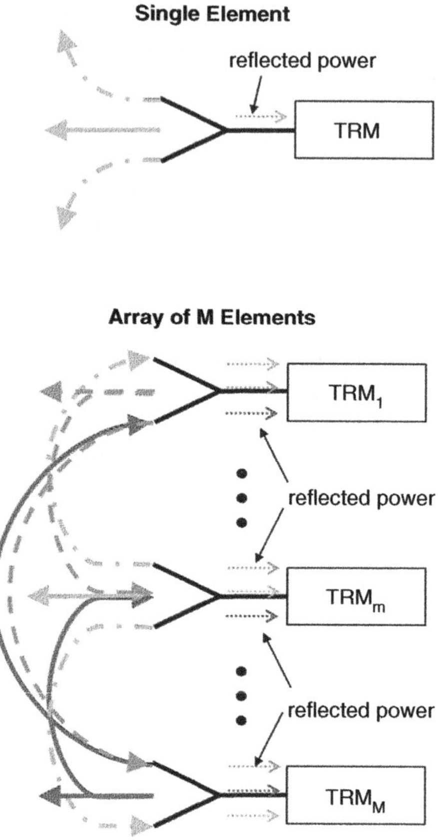

Figure 3.13 Impedance match of a single isolated element is different than the impedance of an array element due to mutual coupling. The impedance match in an AESA is referred to as the active match.

array element match will be different from the isolated element's match because adjacent elements will radiate power via mutual coupling (Pozar, 1994). This is illustrated in Figure 3.13.

Figure 3.14 (Pozar, 1994), shows an example where the element gain of a single array element differs from the elem gain of the *exact same* element in an infinite array of identical elements. In the figure, the array element gain has a null that does not exist in the gain pattern of the isolated element. This is due to mutual coupling interactions in the array environment and is extremely important to characterize for AESA performance. Both (Mailloux, 1993) and (Pozar, 1994) provide excellent detailed derivations for the active match. Fundamentally, the expression for the active match can be obtained by using the scattering matrix concept from microwave network analysis. In an N-port microwave network, the incident

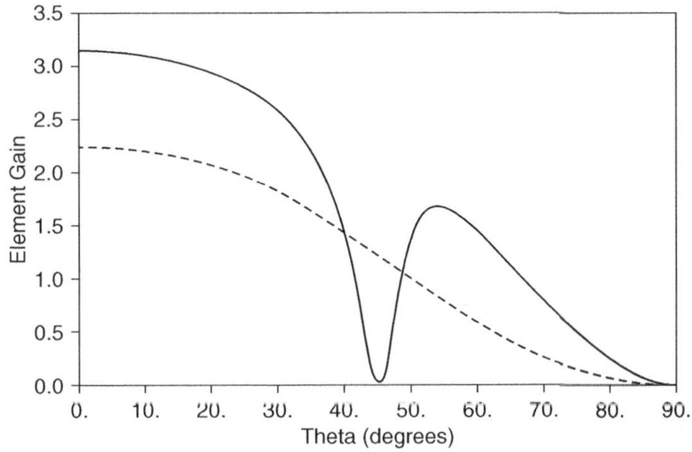

Figure 3.14 The dashed line is the gain pattern for array elem gain in an infinite array. The solid line is an isolated element with the same design that does not have the same gain response because it is not affected by mutual coupling of neighboring elements.

voltage at a port is the sum of the transfer response from all other (N – 1) ports. Mathematically, this can be expressed as (Pozar, 2012):

$$\begin{bmatrix} V_1^- \\ V_2^- \\ \vdots \\ V_N^- \end{bmatrix} = \begin{bmatrix} S_{11} & S_{12} & \cdots & S_{1N} \\ S_{21} & & & \vdots \\ \vdots & & & \\ S_{N1} & \cdots & & S_{NN} \end{bmatrix} = \begin{bmatrix} V_1^+ \\ V_2^+ \\ \vdots \\ V_N^+ \end{bmatrix} \tag{3.45}$$

where

$$S_{mn} = \left. \frac{V_m^-}{V_n^+} \right|_{V_k^+=0 \ for \ k \neq n} . \tag{3.46}$$

In Equations 3.45 and 3.46, V_i^- is the reflected voltage at port m, V_n^+ is the incident voltage from port n, and S_{mn} is the S-parameter, which and is the voltage reflected at port m from port n when all other ports not equal to j are matched. This is a very useful and powerful expression because it mimics exactly how an N-port microwave device is measured with a network analyzer. By measuring the S-parameters, the reflection coefficient can be characterized for every port. For an array element in an AESA, this same approach can be used to express the active match for every element (Pozar, 1994).

Using the scattering matrix in Equation 3.45, the reflected voltage at an element in the AESA can be expressed as

$$V_m^- = \sum_{n=1}^{N} S_{mn} V_n^+ . \tag{3.47}$$

Substituting the amplitude and phase applied to an element into V_n^+ gives the following expression

$$V_n^+ = a_n e^{j\frac{2\pi}{\lambda} n\ d\ sin\theta_o}. \tag{3.48}$$

Solving for Γ using Equations 3.47 and 3.48 results in

$$\Gamma_m(f, \theta_o) = \frac{V_m^-}{V_m^+} = \sum_{n=1}^{N} \frac{a_n}{a_m} S_{mn} e^{j\frac{2\pi}{\lambda}(m-n)\ d\ sin\theta_o} \tag{3.49}$$

and is the expression for the active match for an array of single-port elements.

The utility of Equation 3.49 is that by measuring S_{mn} for all of the array elements, the active match can be calculated. In practice this is done by fabricating a partial array, which is a subset of the number of elements in the AESA. The number of elements is selected to be such that the partial array is 5 λ in any dimension based on (Holter and Steyskal, 2002). The S-parameters are then measured by loading all of the elements except for two. One element is excited using one port of a network analyzer, and the other element is used to measure the received (reflected) energy using the other network analyzer port. This measurement is repeated $N_{partial} - 1$ times while keeping the excited port the same. Once completed, the excited port is changed to another port, and the process is repeated. Although tedious, once complete, the active match for all the elements can be analyzed for quantifying performance. For large arrays, the active match for a single element can be used to represent the match of all elements in the array. For smaller arrays, edge elements are significantly different electrically because they are not surrounded by a full complement of neighboring elements like those near the center of the array. In this case, the active match for one array element cannot be used to represent all elements and must be accounted for in the design.

3.7 Scan Loss

Scan loss is attributed to the loss in main beam energy as the beam is electronically scanned spatially. This directly affects the AESA gain, which in turn affects the transmit *ERP* and receive $\frac{G}{T}$. In Chapter 2, it was shown that the one-dimensional AESA pattern can be computed using pattern multiplication:

$$F(\theta) = EP \cdot AF = cos^{\frac{EF}{2}}\theta \cdot \sum_{m=1}^{M} a_m e^{j(\frac{2\pi}{\lambda}x_m sin\theta - \frac{2\pi}{\lambda_o}x_m sin\theta_o)}. \tag{3.50}$$

The one-dimensional expression can be used to describe the effects of scan loss for simplicity. The same principle applies generally for the two-dimensional expression as well. In Equation 3.50, a cosine expression is used to model the spatial variance of a single-element pattern. In practice, this works very well. Element patterns are measured in an antenna range and values for the element factor, *EF*, are computed using a curve fit to the measured data. A well-designed array element will have an *EF* of 1.1 to 1.2 for AESAs with a small operational bandwidth in the 100s of MHz. For large operational bandwidths, ~ 500 MHz, the EF will vary from 1.2 to 2. At the system level, an *EF* of 2 is usually used to provide margin for initial system analysis and then modified as modeled and measured results become available.

Figure 3.15 provides a graphical representation of the pattern multiplication described in Equation 3.50. In Figure 3.15 the element pattern has a peak at boresite, so the overall AESA pattern suffers no scan loss. The figure also illustrates that the AESA can be scanned to ±40° with less than 1 dB of loss due to the element pattern roll-off. This means that for applications where the AESA required scan is less than 45°, the array element design will not be as challenging.

For most modern AESAs, the scan requirement based on the required field of view (FOV) is generally greater than 45° and can be as large as 70° in some instances. This becomes very challenging for the array element designer as he has to work closer to the limits of physics. This is demonstrated in Figure 3.16 showing an AESA scan to 60°. The AESA suffers ~ 5 dB of scan loss at 60° and increases rapidly beyond that. At the system level, this means that for transmit, more radiated power is required to offset the loss in gain, and for receive the aperture must be made larger and/or the noise figure improved. A methodology

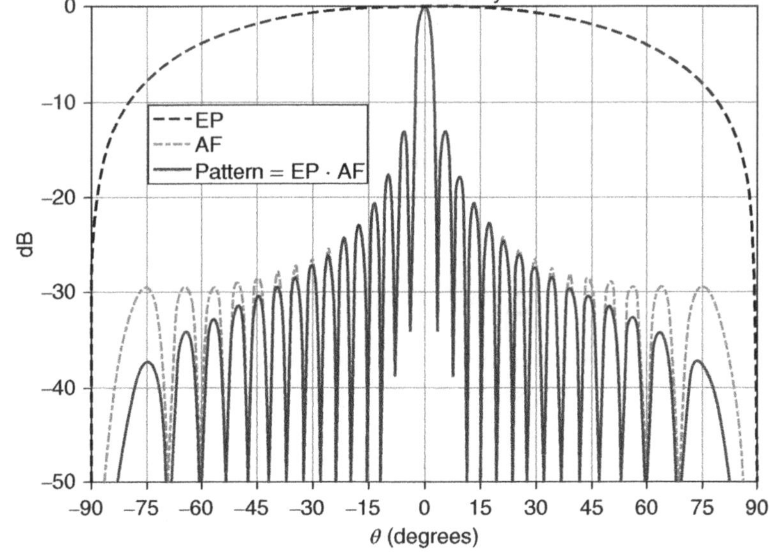

Figure 3.15 Pattern multiplication - boresite (no scan), EF = 1.5.

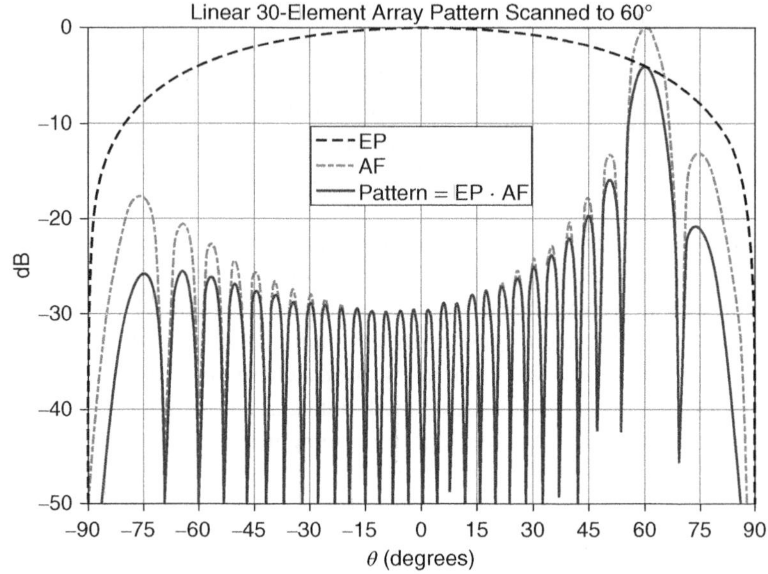

Figure 3.16 Pattern multiplication - 60° scan, EF = 1.5.

for reducing the loss at the maximum scan angle is to design the element pattern to have a peak away from boresite. This introduces loss at boresite but minimizes the loss at the max scan angle. Even with clever techniques such as this, designing an array element that has minimum scan loss over multiple gigahertz of bandwidth and over large scan angles is very challenging.

In describing scan loss above, it was assumed that the array was electrically large (> 5 lambda (Holter and Steyskal, 2002)). However, for cases where the AESA is electrically small (< 5 lambda), or non-planar, pattern multiplication no longer applies and the contributions from each element pattern must be considered as they will be different spatially.

$$F(\theta) = \sum_{m=1}^{M} cos^{\frac{(EF)m}{2}}\theta \cdot a_m e^{j(\frac{2\pi}{\lambda}x_m sin\theta - \frac{2\pi}{\lambda_o}x_m sin\theta_o)}. \tag{3.51}$$

Equation 3.51 still shows that the resultant scan loss will be the sum of the individual element patterns. By setting $\theta = \theta_o$, Equation 3.51 becomes

$$F(\theta) = \sum_{m=1}^{M} cos^{\frac{(EF)m}{2}}\theta \cdot a_m, \tag{3.52}$$

and the resulting pattern is reduced to the sum of the element patterns and any weighting that is applied to the array elements. The impact of this is that analytically the entire array must be modeled and not just a single element with periodic boundary conditions, and/or each individual element pattern must be measured. In Equation 3.52, a_m is an amplitude weight. In some cases it is necessary to apply complex weights B_m to each element pattern where $B_m = a_m e^{j\phi_m}$, and ϕ_m is the phase of the complex weight B_m. The complex weights are used to maximize the sum of the element pattern energy by creating a peak coherent sum at the desired scan angle. This results in a modified expression for Equation 3.52:

$$F(\theta) = \sum_{m=1}^{M} cos^{\frac{(EF)m}{2}} \theta \cdot a_m e^{j\phi_m}. \tag{3.53}$$

References

Balanis, C. *Antenna Theory Analysis and Design*. John Wiley & Sons, Publishers, Inc., 1982.

Holter, H., and Steyskal, H. "On the size requirement for finite phased-array models." *IEEE Transactions on Antennas and Propagation*, pp. 836–840, 2002.

Ludwig, A.C. "The definition of cross polarization." *IEEE Transactions on Antennas and Propagation*, pp. 116–119, 1973.

Mailloux, R.J. *Phased Array Antenna Handbook*. Artech House, Norwood, MA, 1993.

Pettai, R. *Noise in Receiving Systems*. John Wiley & Sons, 1984.

Pozar, D.M. "The active element pattern." *IEEE Transactions on Antennas and Propagation*, pp. 1176–1178, 1994.

Pozar, D.M. *Microwave Engineering*. John Wiley & Sons, Inc., 2012.

Skolnik, M.I. *Radar Handbook*. McGraw Hill, 1990.

Skolnik, M.I. *Introduction to Radar Systems*. McGraw Hill, 2001.

Stutzman, W.L. *Polarization in Electromagnetic Systems*. Artech House, 1993.

4

Transmit Receive Modules

Key Concepts

- TRM Topologies
- Amplifier Efficiency and Classes
- P_{1dB}, Intercept Point, and Noise Factor
- Wideband Operation
- Spurious Beams
- Probability of Failure and MTBF

4.1 Overview

Chapter 3 discussed the array elements, which serve as the interface between the AESA and free space. The array elements enable the AESA to transmit and receive power spatially over the required frequency range of the AESA system. In order to electronically steer the beams for scanning, more is required than just the array elements. A mechanism for applying the correct phase relationship from element to element is needed. This enables the power radiated/received through the array elements to be coherently added together to produce an array beam. This is accomplished by transmit receive modules (TRMs), which are highlighted in the top level system block diagram shown in Figure 4.1.

The voltage that is distributed on transmit by the beamformer (combined on receive) has a phase applied to it that corresponds to a commanded scan direction. As discussed in Chapter 2, that phase for a 1D array is $-\frac{2\pi}{\lambda_o}x_m sin\theta_o$. In an AESA that requires true time delay for steering, the phase shifter is replaced by a time delay unit.

Active Electronically Scanned Arrays: Fundamentals and Applications, First Edition. Arik D. Brown.

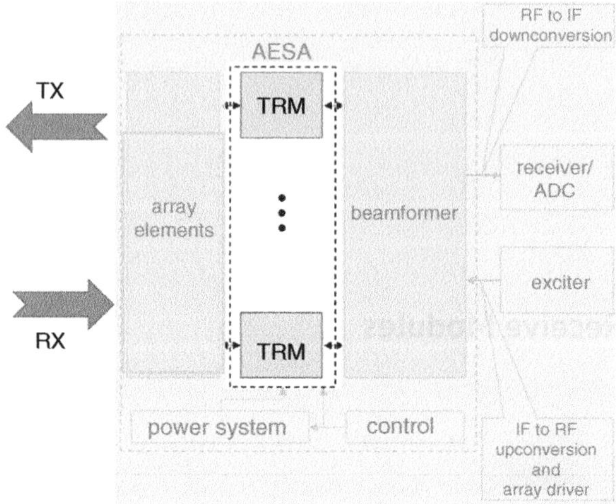

Figure 4.1 The transmit receive modules (TRMs) enable an AESA to electronically scan array beams in addition to power amplification and signal conditioning. They are a critical component in every AESA and typically are the largest cost contributor.

It is important to highlight that the phase shift given to the elements by the TRM is applied to voltages distributed by the beamformer. The TRM applies amplification to the phase modulated signal. The same applies on receive to the voltage received from the array element. The power distributed through the RF chain is the square of the voltage magnitude. This is an important concept for understanding linearity, discussed later, and also how voltage signals are distributed and combined in the beamformer (Chapter 5).

Before exploring the TRM block diagram, it is worthwhile to review mathematically why the TRMs are required to electronically steer the AESA beam. In Chapter 2, it was shown that the pattern expression for an electrically large 1D AESA ($L > 5\lambda$ (Holter and Steyskal, 2002)) is:

$$F(\theta) = cos^{\frac{EF}{2}}\theta \cdot \sum_{m=1}^{M} A_m e^{j\frac{2\pi}{\lambda_o}x_m sin\theta}. \tag{4.1}$$

Setting the complex weight A_m equal to one in Equation 4.1 produces the pattern shown in Figure 4.2. Equation 4.1 shows that when $A_m = 1$, the AESA pattern can only have a maximum value when $\theta = 0^o$.

When an AESA is tuned in an antenna range, the first step is generating a pattern at boresite because no phase shift is required. There are phase shifter and attenuator adjustments applied to balance the phase and amplitude distribution across the AESA due to element errors; however, they are not for steering the beam electronically. By setting the amplitude of the complex weight to one ($|A_m| = 1$) and setting

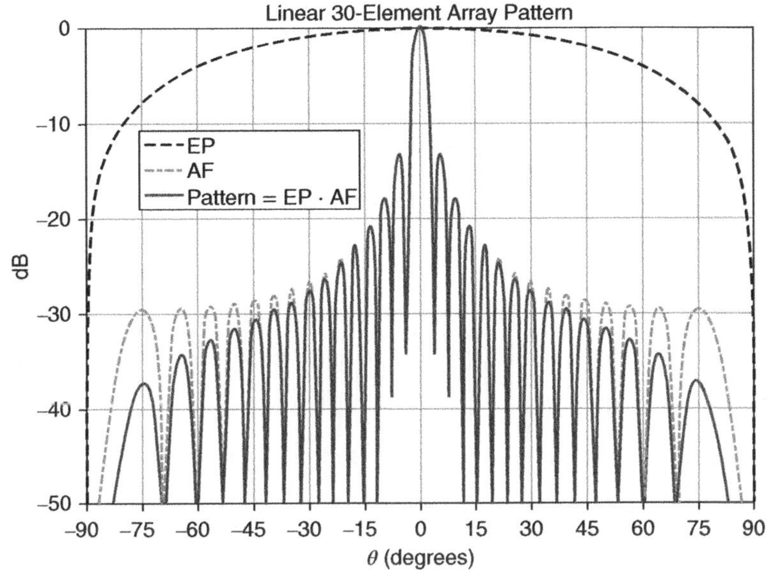

Figure 4.2 When $A_m = 1$, the AESA forms a beam at boresite ($\theta = 0^o$). This is equivalent to not having any phase shifters. Without the complex weight A_m that has the required phase shift $-\frac{2\pi}{\lambda_o} x_m sin\theta_o$, the AESA beam cannot be steered electronically.

the phase of A_m equal to $-\frac{2\pi}{\lambda_o} md\ sin\theta_o$ the AESA can be steered to any angle θ_o. Figure 4.3 illustrates this for the case where θ_o is set to 60°. This is one of the critical functions that the TRM performs enabling the AESA to be electronically scanned.

In addition to providing phase (or time) delay, the TRM also provides other required capability. This primarily includes transmit/receive isolation, signal conditioning, power amplification, receiver protection, and attenuation. These capabilities are given a short description in Table 4.1 and will be elaborated on when describing the TRM block diagram in Figure 4.4.

Figure 4.4 is a block diagram of a typical TRM with the primary RF components that are contained within it. This chapter will focus on the RF performance characteristics of the module and not the digital aspects. However, it is worthwhile to briefly discuss the importance of command and control of a TRM. Digital circuitry is included in the TRM to provide the ability to control the device and change the phase and gain settings during operation. As an example, in a radar all TRMs must be:

- Able to be commanded to ON, OFF, and STANDBY states;
- Commanded to the correct phase and gain settings for every configured processing interval (CPI);

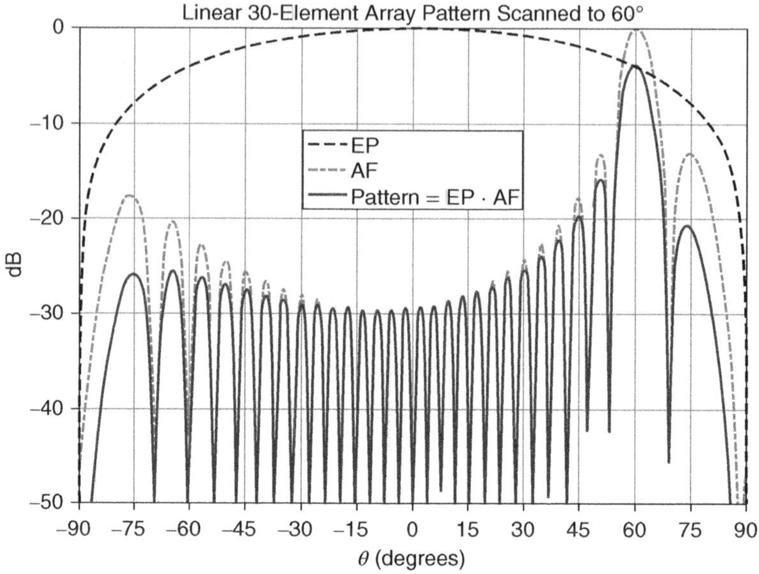

Figure 4.3 When $A_m = 1 \cdot e^{-j\frac{2\pi}{\lambda_0} md\, sin\theta_o}$, the AESA forms a beam at θ_o. This is enabled by the TRM, which is why the TRM is a critical component in the AESA architecture. In the figure, θ_o is equal to 60°.

Table 4.1 In addition to providing the correct phase and/or time delay to each element in the AESA, TRMs also provide transmit/receive isolation, signal conditioning, power amplification, receiver protection, and attenuation

TRM Capability	Description
Transmit/receive isolation	TR switches and circulators provide isolation between transmit and receive operation. This is critical especially in high power transmit AESAs where the amplified transmit power can damage the receive electronics.
Signal conditioning	Filtering is required for linearity on both transmit and receive, in addition to any channelization that is done particularly for wide band systems
Power amplification	Amplifies the signal on transmit to achieve the level required for meeting ERP. Also, low noise power amplification is required on receive.
Receiver protection	RF circuitry is required to prevent internally and externally generated power from damaging the receive electronics.
Gain balance	Used to balance out element to element amplitude errors, in addition to balancing the gain for the RF electronics chain.

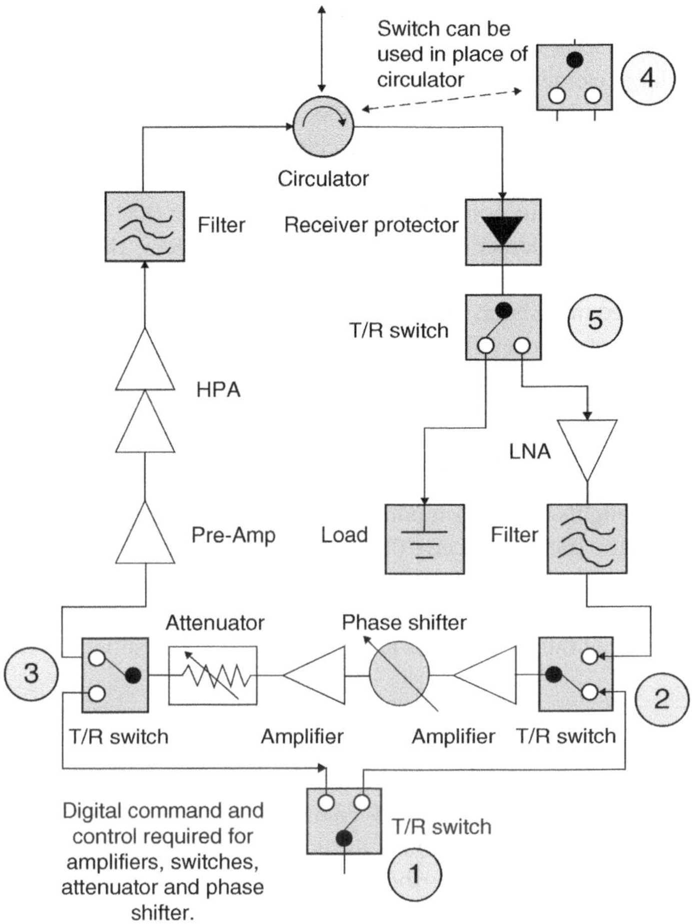

Figure 4.4 The key components of a TRM are amplifiers, switches, filters, phase shifters, attenuators, and receiver protectors with digital control. The switches in the figure are configured for transmit operation.

- Switched repeatedly back and forth between transmit and receive for every pulse transmitted for a pulse radar;
- Able to provide frequent update messages on the TRM health status;
- Aligned with a system level timing strobe;
- Able to be commanded to a built-in-test (BIT) state; and
- Equipped with enough memory to store data tables for accessing phase and gain tables that are a function of frequency and scan angle.

A similar list of related capabilities is required for other non-radar applications.

In addition to the features just listed, the distribution of the command and control to the TRMs is also a key factor in the design of an AESA. For AESAs that have large element counts with 100s to 1000s of elements, the distribution of the command and control to the TRMs can be challenging and requires a well-thought-out digital distribution architecture. As AESAs have advanced, TRMs have also advanced by increasing the number of channels integrated into a single TRM. Up to this point, the assumption has been that each TRM services one channel (array element); however, for large AESAs, the TRMs typically contain multiple channels for a higher level of integration and reduced cost. This will be discussed more later. The increased number of channels per TRM reduces the complexity of the overall digital distribution of the command and control to the TRMs for system operation.

4.1.1 TRM Baseline Topology

Figure 4.4 highlights the key components in a TRM. Although some of the components such as the transmit receive (TR) switches are simple, they still play a vital role in the proper operation of the TRM. This section will provide a description of purpose and function of each of the components shown in Figure 4.4 and conclude with a mapping of the components to the TRM capabilities shown in Table 4.1, in Table 4.2.

4.1.1.1 TR Switches

There are various switches in the TRM. Their purpose is to provide isolation between the transmit and receive RF chains in the AESA and to route the RF voltage along the proper RF chain during operation. In Figure 4.4, five TR switches are shown and are described below:

Table 4.2 TRM components mapped to the capabilities listed in Table 4.1

TRM Capability	TRSW	Amp	Pre-Amp /HPA	LNA	PhsShf	Attn	Filter	Circ	RP
Transmit/ receive isolation	✓							✓	
Signal cond.					✓	✓	✓		
Power amp.		✓	✓	✓					
Receiver protection	✓								✓
Gain bal.						✓			

- TR switches 1-3: Although these switches do provide isolation, their main function is to enable the phase shifters and attenuators to be shared between the transmit and receive paths. Instead of using a separate phase shifter and attenuator for both transmit and receive, the switches allow them to be shared, reducing both the space required for the TRM and the complexity. Additionally, this *common leg* architecture provides the same single tone and two-tone third-order intercept (TOI) performance as a module architecture employing independent phase shifter and attenuator for transmit and receive (Agrawal et al., 1999), (McQuiddy Jr. et al., 1991). Finally, these switches also help to minimize the number of control lines required to command the TRM compared to a TRM that has independent phase shifters and attenuators for transmit and receive.
- TR switch 4: Historically, a circulator has been used to isolate the transmit and receive paths. Increased fidelity in the modeling of the mutual coupling between array elements has enabled TR switches to be used optionally in place of a circulator. This switch has the same function as TR switches 1-3 but has to survive under much higher power levels because it is located after the HPA. As an example, if on transmit a 0.1 W signal is sent to the pre-amp and HPA, which have a combined gain of 30 dB, TR switch 4 will see 100 W of power. Additionally, if the AESA is in a signal dense environment with strong intentional and unintentional interference, TR switch 4 is also susceptible to high power levels on receive. Lastly, this switch requires very good isolation so that high power is not coupled into the receive chain.
- TR switch 5: This switch's purpose is solely to protect the LNA and any sensitive electronics prior to the receiver. When the AESA is switched into transmit, this path is switched to a load that is able to handle high power. This switch serves as a backup to increase the transmit to receive isolation.

4.1.1.2 Amplifiers

Several different types of amplifiers are required in the TRM. The amplifiers referred to here are general amplifiers that are used to balance the gain in the RF chain and also maintain the noise figure on receive. Two general amplifiers are shown in Figure 4.4 and are located in front of both the phase shifter and attenuator. They provide amplification on the shared phase shifter/attenuator string allowing for minimal loss of signal on transmit before the preamplifier and HPA and also prevent the noise figure from degrading due to the losses in the phase shifter and attenuator.

4.1.1.3 Pre-Amplifier and HPA

The preamplifier and high power amplifier (HPA) have the primary responsibility of increasing the transmit signal power to the required level per element to meet the ERP requirement. The amplifiers are required to be linear and have excellent P_{1dB} performance (the amount of power output from the amplifier when it begins to compress). The choice of semiconductor material is important for these amplifiers to ensure that the high gain can be achieved at temperatures that are not excessive due to thermal heating. Mechanically, these amplifiers must be designed with a very good heat sink within the module and in many cases use liquid cooling for heat transfer. As AESA systems become more wideband, the HPA must also support wider band operation. This can be challenging because nonlinear signals can be generated in the HPA, which is disadvantageous.

4.1.1.4 LNA

The low noise amplifier (LNA) serves an equally critical role in receive, similar to the HPA in transmit. The LNA must have a low noise figure with high gain. As discussed in Chapter 3, the first amplifier in an RF chain establishes the noise figure performance for the entire receive electronics chain. Great care is taken when designing the LNA or selecting a COTS part because the $\frac{G}{T}$ performance of the AESA system is dependent on the LNA.

4.1.1.5 Phase Shifter

As previously discussed, the phase shifter enables the AESA to steer the array beam electronically by applying the appropriate phase shift for each AESA element. Ideally, the phase applied by the phase shifter would be smooth and continuous, but in practice it is quantized. The phase shifter is comprised of N bits, or phase states, that approximate the required phase for steering. As an example, a 6 bit phase shifter would have 2^6 phase states with an LSB of 5.625°. In most applications, as shown in Chapter 2, 5-6 bits is sufficient for good pattern performance.

4.1.1.6 Attenuator

Attenuators have various functions that are related to gain control in the AESA. If the gain of the TRM is too high, the attenuator can be used to reduce the signal level and maintain a balanced RF chain. Typically, this has to be done over frequency and temperature to account for gain variations. In addition to the RF chain, the attenuator can be used to provide an amplitude taper across the array to achieve low sidelobe levels. This is typically not the preferred method as it incurs additional loss and reduces the array gain efficiency. Lastly, the attenuators are used to minimize channel-to-channel errors across the array for tuning the array and minimizing undesired SLs due to AESA errors.

4.1.1.7 Circulator

Similar to the TR switches, a circulator is typically used for isolation between the transmit and receive paths. It is required to support the operational bandwidth of the array elements, handle the high power levels that are generated by the HPA, and have low loss. Because of its location in the TRM, the circulator is one of the primary loss sources post-HPA and pre-LNA. After the HPA there is no opportunity to apply additional gain to compensate for the circulator loss, and the same is true on receive in terms of its location to the LNA. As numerical modeling of arrays has improved, current designs are pushing toward removing the circulator from the front end (replacing with a T/R switch) and designing the array element to minimize impedance mismatches between the TRM and array elements.

4.1.1.8 Receiver Protector

The receiver protector (RP) performs exactly what its name implies. It protects the receive chain from high power levels that could be damaging. The receiver protector is designed to stop any power above a certain prescribed level from entering the electronics chain after the array elements. Similar to the circulator, it has a disadvantageous location of being located pre-LNA. Its loss is a direct 1:1 dB addition to the noise figure.

4.1.1.9 Filters

Filters provide several different functions in the TRM. They are used to filter the HPA output prior to transmission from the array elements to minimize undesired spurious products generated by the HPA. Filters are also used on transmit and receive to provide channelization over a large bandwidth that could be an octave (twice the lowest frequency) or greater. The filters for channelization can be tunable or switchable. Tunable filters, although more difficult to be built, have the advantage of not being limited by band breaks in comparison to a switchable filter.

4.1.2 TRM Topology Types

Figure 4.4, shows the functionality of a standard TRM. It has transmit and receive paths without any special filtering or integration. This TRM topology supports both radar and EW. For EW, this TRM would be capable of supporting both EA and ESM. In this section, several alternative TRM topologies will be covered that are specific to the type of mission that the AESA is required to support.

4.1.2.1 Receive Only

In receive-only applications such as ESM or SIGINT, a transmit channel is not required. This simplifies the TRM, since only a receive channel is required. Figure 4.5 shows a receive only TRM block diagram. Generally, a receive-only AESA typically is used for large operational bandwidths and requires some level of channelization. Additionally, this type of TRM must support multiple simultaneous beams, which adds to the TRM electronics density.

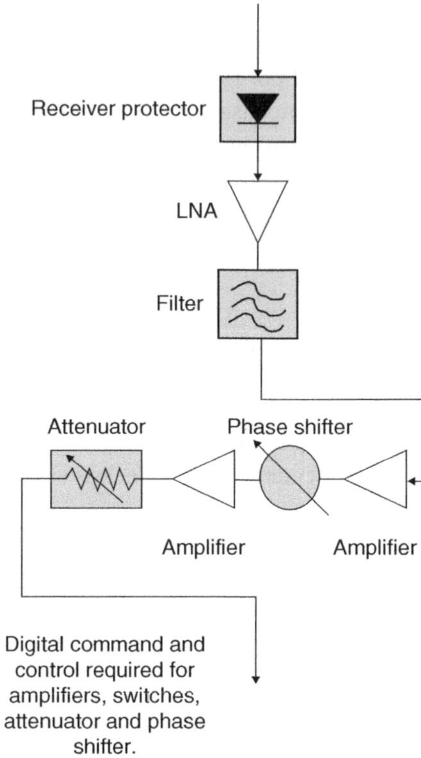

Figure 4.5 A receive-only TRM does not require a transmit channel. Also, receive-only AESAs are typically used for wide bandwidth applications where channelization is required in addition to multiple beams.

4.1.2.2 Channelization

For systems that have large operational bandwidths that span more than an octave, channelization is typically required. This places a requirement on the TRM to operate over a wide bandwidth. This is done with channelized filters. Figure 4.6 shows a TRM that has filters for channelization in the receive path. (The same can be done similarly in the transmit path.) Tunable filters are an option as well since they do not have the limitation of band breaks that the channelized filter has. The AESA can be tuned to any frequency. Tunable filters tend to be more challenging than channelized filters, but with the advancement of filter technology, they are becoming more tenable for use.

Channelized filters in the TRM also provide additional linearity performance. Higher-order nonlinear signals are generated by wideband amplifiers. The filter enables undesired nonlinear signals to be filtered out prior to the receiver and on transmit prior to the HPA.

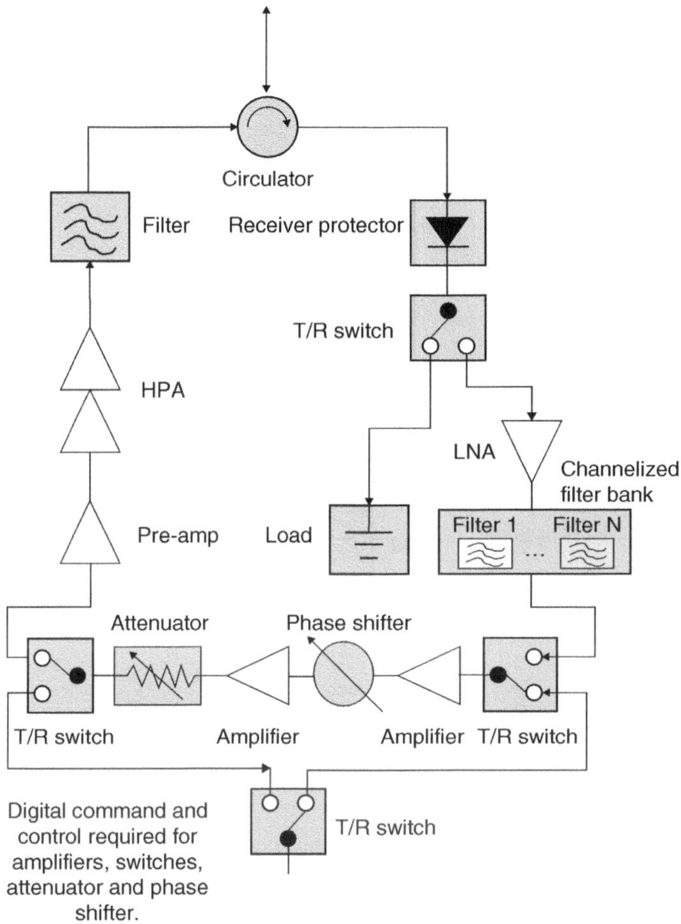

Figure 4.6 TRM channelization stops out of band nonlinear signals generated by the amplifier from being passed to the receiver.

4.1.2.3 Simultaneous Beams

In applications that require multiple simultaneous receive beams, additional RF circuitry is required to support simultaneous beam operation. Figure 4.7 shows the the additional circuitry in the TRM. A multiplexer is required from the primary wide bandwidth LNA output. Each leg of the multiplexer feeds an independent electronics chain. This adds complexity to the TRM, and care must to be taken to minimize the additional RF circuitry required.

4.1.2.4 Multi-Channel TRMs

As was previously discussed, TRMs are the largest cost contributor to an AESA and can be as much as 40–60% of the entire AESA cost (Agrawal et al., 1999). This

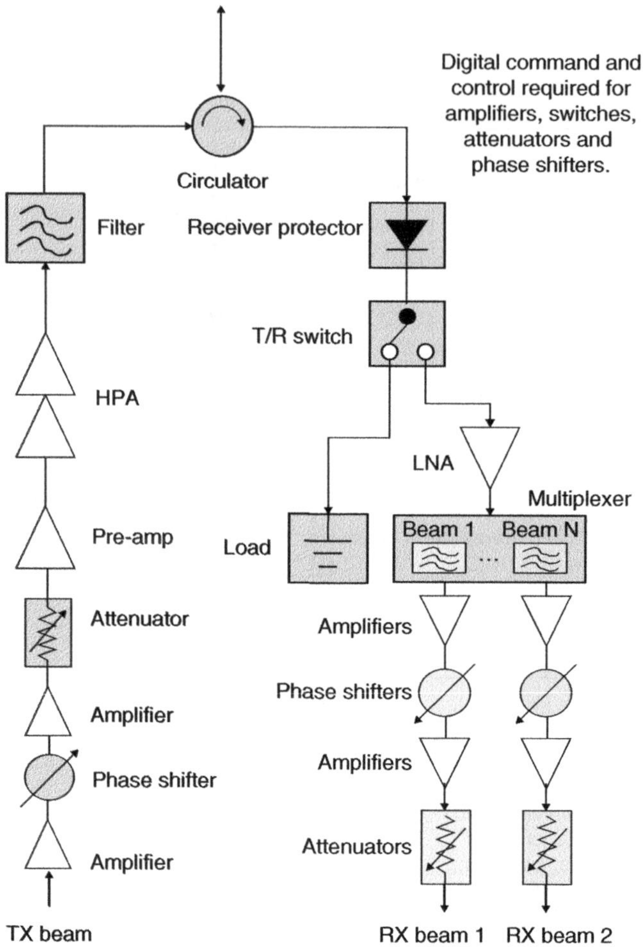

Figure 4.7 Multiplexer and additional channel paths with independent phase shifters and attenuators enables the TRM to support simultaneous AESA beams.

is because a TRM is required for every array element. In order to minimize this cost, groups of channels are integrated into the same package, which results in a multi-channel TRM. This TRM topology is shown in Figure 4.8. As an example of the advantage of integrating multiple channels in the same TRM mechanical package, consider a 64 element (8 × 8) AESA. Without a multi-channel TRM, this would require 64 TRMs. However, by combining 4 elements into a single TRM, only 16 TRMs would be required so that each support 4 elements. This helps to decrease the cost and also simplifies the interconnects required for the beamformer.

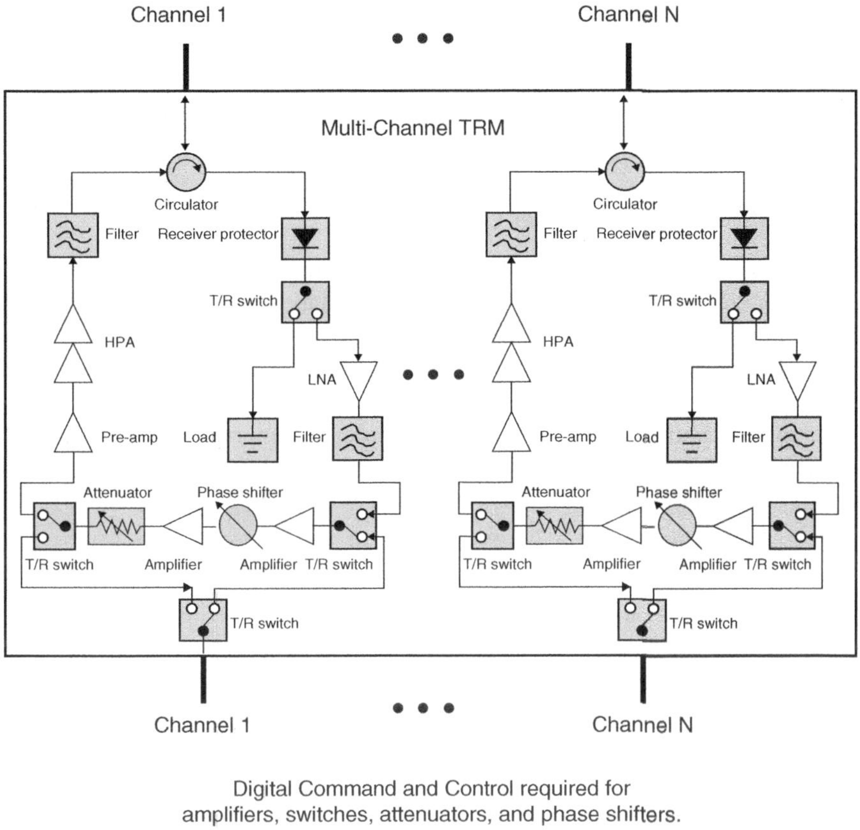

Figure 4.8 AESAs with a large number of elements employ multi-channel TRMs to simplify the amount of digital control required, in addition to reducing the size, weight, and cost of the AESA.

4.2 Transmit Operation

The primary purpose of the TRM during transmit operation is to amplify the signal to a level that will satisfy the ERP requirement for the AESA. This is accomplished by the HPA, which is typically preceded by a preamplification stage as shown in Figure 4.4. The HPA will nominally have a gain value of 20 dB (factor of 100) or more. For AESAs, the HPAs are typically MMIC devices that are designed into a heat sink that is built into the TRM package design. In the early development and use of AESAs, gallium arsenide (GaAs) was used as the primary semiconductor for HPA fabrication. A key attribute of GaAs is that it has very high electrical resistivity, which provides low attenuation (loss) for RF signals (Walker, 2012).

For applications requiring greater than 10 W per element and as high as 100 W per element gallium nitride (GaN) has become the industry standard. GaAs is not suited well for high-power applications (Walker, 2012); however, GaN is able to provide high power with excellent efficiency. GaN HPAs are now commonly used in radars and EA systems where high ERP performance is required.

4.2.1 Efficiency and Amplifier Classes

Amplifiers are categorized in alphabetical classes that define their performance parameters for selection in applications. The alphabetical classification of electronic amplifiers dates back to the earliest era of electronics (Walker, 2012). Before describing the various classes, it is important to define two of the figures of merit for an amplifier, which are efficiency and power added efficiency (PAE). An amplfier requires a DC bias voltage in order to generate RF output power. Efficiency, η, describes the relationship between the DC input power and the RF output power of the amplifier and is expressed as (Walker, 2012), (Pozar, 2012):

$$\eta = \frac{P_{out}}{P_{DC}}, \tag{4.2}$$

where P_{out} is the RF output power of the amplifier, and P_{DC} is the DC input power. Amplifiers typically are more efficient as more DC input power is applied. Typically, the TRM has a DC power requirement and an output RF power requirement that dictates the class of amplifier that can be used for the TRM HPA.

Another useful figure of merit for an amplifier is PAE and is expressed as

$$\text{PAE} = \frac{P_{out} - P_{in}}{P_{DC}}. \tag{4.3}$$

As previously mentioned, the output RF power of the HPA can be one hundred (20 dB of gain) to one thousand (30 dB of gain) times larger than the input RF power. Equation 4.2 does not consider the input RF power, and Equation 4.3 better quantifies the amplifier's performance. Equation 4.3 can further be modified as (Pozar, 2012):

$$\text{PAE} = \frac{P_{out} - P_{in}}{P_{DC}} = \left(1 - \frac{1}{G}\right)\frac{P_{out}}{P_{DC}} = \left(1 - \frac{1}{G}\right)\eta \tag{4.4}$$

where G is the power gain of the amplifier.

Having established the definitions for efficiency, the amplifier classes can be described. The class of an amplifier describes how its input current is biased. Class A amplifiers have very good linearity with a maximum theoretical efficiency of 50% (Walker, 2012). They are biased to conduct over the entire range of the input signal cycle. Most small-signal and low noise amplifiers are class A (Pozar, 2012). Class B amplifiers are biased to their threshold point and draw no current until an input signal is applied (Walker, 2012). Class B amplifiers have a theoretical maximum efficiency of 78% (Walker, 2012). Class AB amplifiers are similar to class A amplifiers except that they are biased at a lower value. Class C amplifiers can achieve higher efficiencies than classes A, AB, and B but are only suited for applications that use constant amplitude modulation. Classes D, E, F, J, and S are used to increase the efficiencies of the previously mentioned classes and are more for niche applications.

4.2.2 P_{1dB}

Amplifiers are used to generate an output power that is a scaled version of their input power. This is shown in Figure 4.9. Ideally, it is desired for the output power to increase linearly as a function of increased input power. Because amplifiers are nonlinear devices, the ideal scenario does not apply. The amplifier generates nonlinear components that must be accounted for in its design.

Figure 4.10, shows a representative nonlinear amplifier response that shows the output power as a function of the input power. There is a range of input power values where the output power is a linear function of the input power.

This linear function is the gain of the device. This linear region is typically referred to as the small-signal response of the amplifier. As the power of the

Figure 4.9 An amplifier scales its input power for an increased output power. The scaling factor is the gain of the amplifier, and the output power is equal to the gain multiplied by the input power.

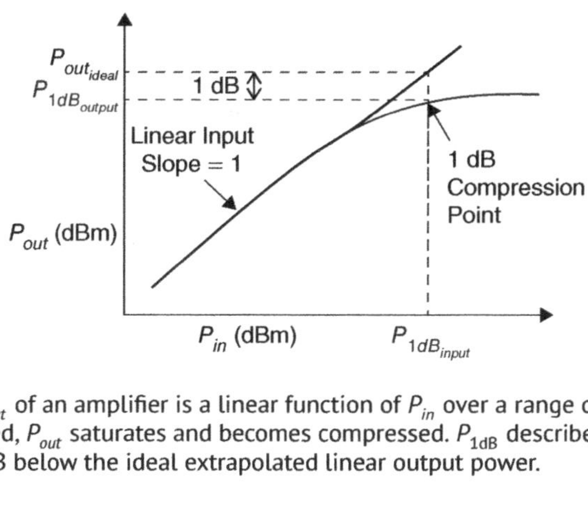

Figure 4.10 P_{out} of an amplifier is a linear function of P_{in} over a range of input powers. As P_{in} is increased, P_{out} saturates and becomes compressed. P_{1dB} describes the point at which P_{out} is 1 dB below the ideal extrapolated linear output power.

amplifier is increased, the output power does not continue to increase linearly indefinitely. Instead, the output power begins to saturate and rolls off. This characteristic is referred to as compression, and this region is typically referred to as the large-signal response of the amplifier. To characterize the compression behavior at the 1 dB compression point, P_{1dB}, is defined. It is the point in the response where the output power is 1 dB less than the ideal gain slope of the amplifier. This is shown in Figure 4.10. This relation can be expressed as (Pozar, 2012):

$$OP_{1dB} = (IP_{dB} + G) - 1\ dB \tag{4.5}$$

Equation 4.5, is expressed in dB, where *OP* is the output power of the amplifier, *IP* is the input power of the amplifier, and *G* is the amplifier gain. In some instances there are regions in the linear response of the amplifier where the response is not linear. This is referred to as gain compression and provides an additional challenge to the TRM design. The nonlinear region of the linear response must be measured and characterized and can change over temperature and frequency.

For applications requiring high values of *ERP* such as radar and EA, the HPAs of the AESA are operated at/near the P_{1dB} point. This is because the amplifier is most efficient in this region. The output power of the TRM must be characterized over frequency and temperature to ensure that the HPA is not being overcompressed and also that it is being driven hard enough to maximize efficiency. As AESAs have advanced toward multiple simultaneous beams on transmit, this modifies the approach to operating the HPA. The HPA must be operated backed off from compression so that simultaneous signals do not create higher-order products at the HPA output and/or oversaturate the amplifier output. As a rule of thumb the HPA must be backed of by $20 \cdot log_{10}(Number\ of\ input\ signals)$. This can be relaxed by using pre-distortion techniques (Walker, 2012), and/or higher fidelity characterization of the amplifier response.

4.2.3 Linearity

4.2.3.1 Harmonics and Intermodulation Products

Traditionally, when linearity is discussed it is assumed to be referencing operation in receive. However, with the increasing interest in wideband AESAs on transmit, it is applicable for this situation also. As an example, when an AESA is operated near other mission critical assets it is required that the AESA not interfere with communication links. Wideband AESAs can generate harmonics and intermodulation products at the communication link frequencies and cause undesired interference. This is especially true for EA applications where the AESAs can unintentionally jam friendly payloads on the same platform.

To understand how nonlinearities are generated by an amplifier, the response to an input voltage signal is considered. The amplifier response acts on the input voltage, V_{in}, and the corresponding input power is $P_{in} = |V_{in}|^2$. The output voltage of the amplifier can be expressed in a power series as:

$$V_{out} = a_o + a_1 V_{in} + a_2 V_{in}^2 + a_3 V_{in}^3 + \cdots + a_n V_{in}^n \tag{4.6}$$

The a_o term in Equation 4.6 is a DC term, a_1 is the linear gain, and the other a_n terms are the gains of the higher-order products generated by the amplifier. The higher-order products for $n > 3$ are usually not considered because their amplitude is small compared to the fundamental signal ($n = 1$).

To characterize the linearity performance, two types of input are typically considered. They are single-tone and two-tone inputs. For single-tone performance the amplifier input is

$$V_{in} = V_o cos(\omega_1 t). \tag{4.7}$$

Substituting Equation 4.7 into Equation 4.6 leads to

$$V_{out} = a_o + a_1 V_o cos(\omega_1 t) + a_2 V_o^2 cos^2(\omega_1 t) + a_3 V_o^3 cos^3(\omega_1 t) + \cdots \tag{4.8}$$

Using trigonometric identities, the higher order terms can be expanded out to express the higher order output voltages in terms of linear frequency components that are multiples of the fundamental ω_1. A 1 is used as the subscript for the fundamental frequency so that later when the two-tone input is described, the number 2 will be used as the subscript to denote the frequency for the additional input voltage term.

The second-order term in Equation 4.8 will be explained in more detail as an example of how to express the higher-order terms as linear frequency components at the output of the amplifier. Using the trigonometric half-angle identity

$$cos^2(A) = \frac{1 + cos(2A)}{2} \tag{4.9}$$

the second-order term in Equation 4.8 can be expressed as

$$\begin{aligned} a_2 V_o^2 cos^2(\omega_1 t) &= a_2 V_o^2 \frac{1 + cos(2\omega_1 t)}{2} \\ &= \frac{a_2 V_o^2}{2} + \frac{a_2 V_o^2 cos(2\omega_1 t)}{2}. \end{aligned} \tag{4.10}$$

Equation 4.10 shows that the square term in Equation 4.8 generates a harmonic at the amplifier output that is twice the frequency of the fundamental input signal. This holds true for the other higher-order components as well. A single-tone input at the fundamental frequency ω_1 will generate harmonics at the output of the amplifier at the frequencies $2\omega_1, 3\omega_1, \cdots, n\omega_1$. This is illustrated in Figure 4.11. To counteract this on transmit, a filter can be used at the output of the TRM to filter out harmonic outputs. Since the harmonic amplitudes decrease as the order increases, typically only the second- and third-order harmonics are of concern and can be mitigated with a filter.

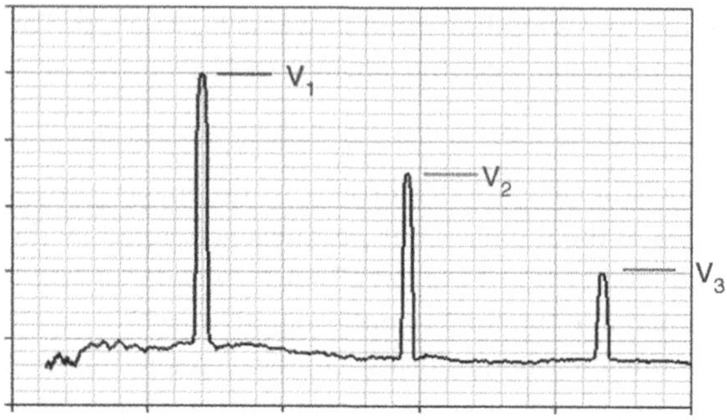

Figure 4.11 The fundamental, second harmonic, and third harmonic are shown. As the order of the harmonic increases, the amplitude decreases.

A two-tone input refers to two signals that are combined and input into the amplifier. This is a very important case because the two-tone case produces intermodulation products (intermods) that in many cases cannot be filtered and must be addressed by high linearity performance from the HPA (on receive the LNA) or by carefully managing the transmit signal chain gain. To properly characterize the linearity performance of the TRM, the two-tone input case must be measured to ensure satisfactory performance from the AESA.

To express the two-tone amplifier output, the input voltage can be expressed as:

$$V_{in} = V_o(cos(\omega_1 t) + cos(\omega_2 t)). \tag{4.11}$$

This assumes that the inputs at different frequencies have the same amplitude, which in practice is not always the case; however, it is sufficient for characterizing performance. Substituting Equation 4.11 into Equation 4.8 leads to

$$\begin{aligned} V_{out} = a_o + a_1 V_o(cos(\omega_1 t) + cos(\omega_2 t)) + a_2 V_o^2(cos(\omega_1 t) + cos(\omega_2 t))^2 + \\ a_3 V_o^3(cos(\omega_1 t) + cos(\omega_2 t))^3 + \cdots . \end{aligned} \tag{4.12}$$

Expanding Equation 4.12 using trigonometric identities produces intermods with frequencies in the form of (Pozar, 2012):

$$p\omega_1 + q\omega_2, \quad p, q = 0, \pm 1, \pm 2, \pm 3, \cdots \tag{4.13}$$

Figure 4.12 illustrates the intermodulation products that are described in Equation 4.13 up to the fourth order.

Most of the intermods shown in Figure 4.12 can be filtered out with a bandpass filter or low pass filter. However, the intermods of the form $2\omega_1 - \omega_2$ and $2\omega_2 - \omega_1$, cannot be easily filtered out. These intermods are the most challenging to deal with in an AESA for narrowband or wideband operation and are typically the largest driver for AESA linearity performance.

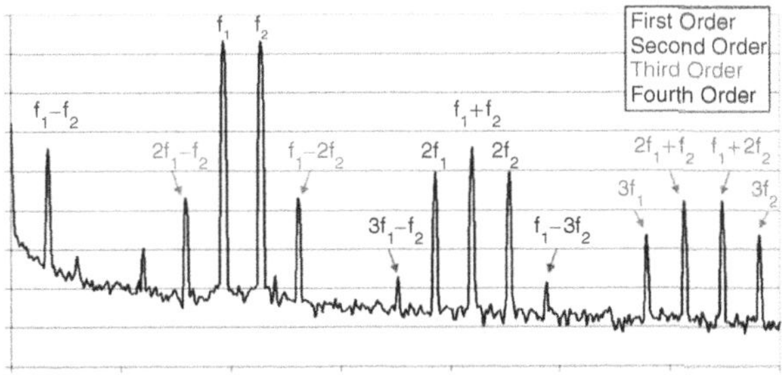

Figure 4.12 Two-tone input into an amplifier generates both harmonics and intermodulation products. This input condition can be very challenging to handle in wideband systems. The TRM is tested extensively to ensure the nonlinear products are at a sufficiently low level relative to the fundamental signal inputs.

4.2.3.2 Intercept Point

The previous section described how intermods are generated in a nonlinear device, specifically the amplifiers in a TRM. Similar to P_{1dB}, there are expressions for the power of the intermods that are expressed in terms of the linear input power. These expressions are called the intercept points. As previously discussed, and also highlighted in Figure 4.12, the intermods with an order greater than three have an amplitude that is much smaller than the fundamental or can be easily filtered out. Because of this, the second-order intercept point (IP_2) and third-order intercept point (IP_3) are primarily considered for evaluating linearity performance.

Figure 4.13 depicts the IP_2 and IP_3 for a nonlinear device. The intercept point is defined as the power level where the output fundamental power is equal to the output of the nth-order power level. This is shown in Figure 4.13 for both IP_2 and IP_3. This definition is based on the ideal slopes of the fundamental and nth-order power; however, as previously discussed and as shown in Figure 4.13, the

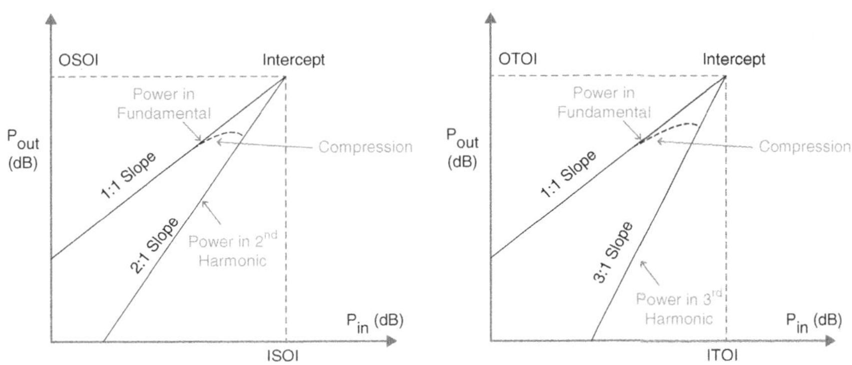

Figure 4.13 Graphical illustration of the definitions for IP_2 and IP_3.

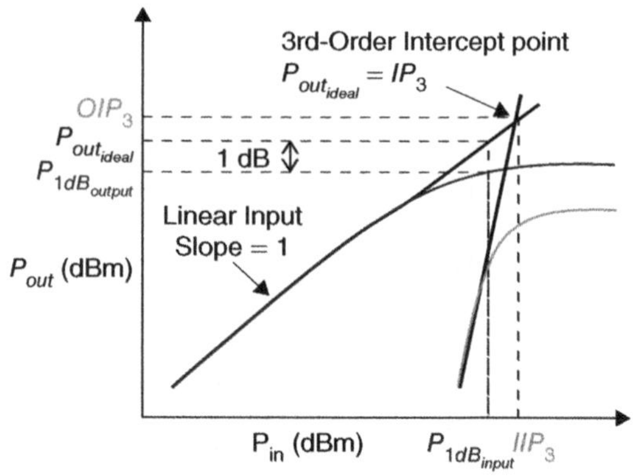

Figure 4.14 The third-order output power (and other *nth*-order output powers) compresses similarly to the fundamental power for P_{1dB}. The intercept definition uses the linear slope of the fundamental and *nth*-order power.

fundamental power output compresses and does not continue linearly indefinitely. The same compression behavior is applicable to the *nth*-order power levels and is illustrated in Figure 4.14 for IP_3 (Pozar, 2012).

The *nth*-order output power can be expressed as a function of the fundamental output power, and the output intercept point and can be written as (see Appendix D):

$$P_{o_n} = nP_{o_1} + (1-n)OIP_n. \tag{4.14}$$

In Equation 4.14, P_{o_n} is the *nth*-order output power, P_{o_1} is the fundamental output power, and OIP is the output intercept point. By rearranging the terms in Equation 4.14, OIP_n can be represented as

$$OIP_n = \frac{1}{(1-n)}P_{o_n} - \frac{n}{(1-n)}P_{o_1}. \tag{4.15}$$

The input intercept point *IIPn* can be related to the OIP_n by

$$OIP_n = G + IIP_n. \tag{4.16}$$

Using Equation 4.14 leads to the following expressions for OIP_2 and OIP_3:

$$\begin{aligned} OIP_2 &= 2P_{o_1} - P_{o_2}, \\ OIP_3 &= \frac{3}{2}P_{o_1} - \frac{1}{2}P_{o_3}. \end{aligned} \tag{4.17}$$

Reexamining Equation 4.14, it is seen that the nonlinear power P_{o_n} is reduced by increasing the *IP*. Higher *IP* correlates to increased linearity performance for a nonlinear device and subsequently the TRMs in an AESA. Additionally, Equation 4.14, can be used to derive the expression for spurious free dynamic range (SFDR) of the AESA. This will be covered in detail in Chapter 6.

4.2.4 Wideband Operation

As the desire for AESAs with increased bandwidth has increased, this has put a requirement on the TRMs to support wider bandwidths that cover multiple octaves. Additionally, transmitting multiple simultaneous beams is a capability that is also being leveraged on AESAs. The implications of wideband and/or simultaneous beams for transmit operation has created the need for filtering after the HPA. The filter shown after the HPA in Figure 4.4 would not be required for a suboctave AESA that only transmits a single beam. This filter, though, only helps to suppress out-of-band harmonics and intermods.

For multioctave operation, the TRM must be able to transmit at any frequency within the operational bandwidth. As an example, consider an AESA that must support an operational bandwidth of 1 GHz to 4 GHz. For the TRM to support this, the HPA must be able to operate from 1GHz to 4 GHz also. Based on the previous discussion on linearity, this means that the amplifier when operating from 1 GHz to 2 GHz will generate in-band and out-of-band harmonics (which can be filtered) from 2 GHz to 4 GHz. The in-band harmonics will affect the RF signature of the system, which may or not be an issue, depending on the application. For the same example, operating at frequencies from 2 GHz to 4 GHz will generate out-of-band harmonics. These nonlinearities can be filtered. This is illustrated in Figure 4.15.

Simultaneous beams exhibit the same phenomenon as multioctave operation. Because multiple signals with different frequencies are being input into the HPA, both harmonics and intermods are created. The filter post-HPA can filter the out-of-band nonlinearities (Figure 4.16); however, it does not remedy the in-band nonlinearities, which must be addressed by the intercept points of the HPA.

4.2.4.1 Nonlinear Beams

The preceding sections discussed the harmonic and intermodulation products that are generated by active nonlinear devices and their intercept points. With the increased prevalence of wideband AESAs, this topic has increased in importance because the HPA and LNA in the TRM must be wideband to support the system and drive both transmit and receive performance. For AESA architectures that support multioctave operational bandwidths, this phenomenon must be modeled and understood to mitigate impacts on system performance. The analysis of HPA generated harmonics being radiated as beams for active phased arrays was initially studied in the 70s (Sandrin, 1973). However, at that time AESAs were not wideband. With the increase in AESA operational bandwidth this has been studied further with examples in (Hemmi, 2002), (Loyka, 2003), and (Chun et al., 2016).

The expressions shown in this section will be based on a 1D AESA with no loss of generality as compared to the 2D case. Expressions can be derived from

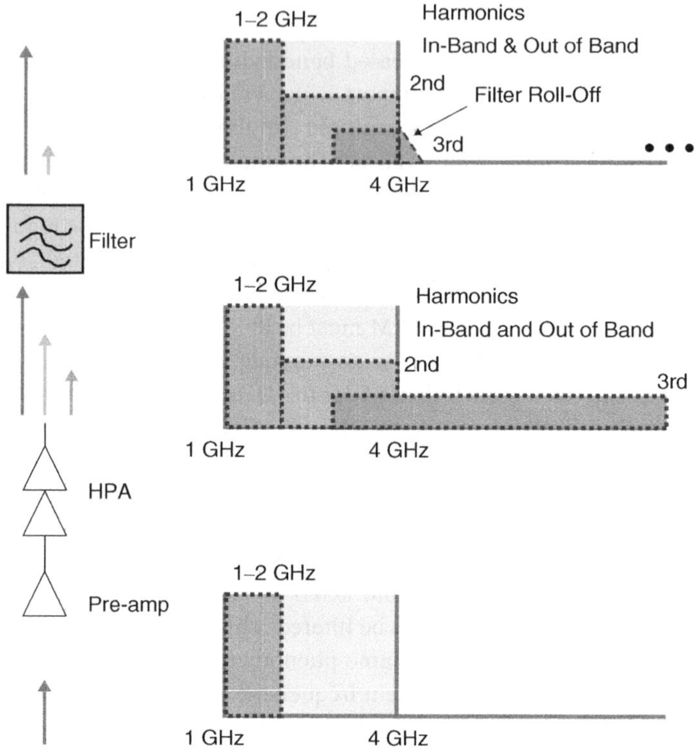

Figure 4.15 Output harmonics generated by in-band signals can be filtered. Due to the finite roll-off of the filter, a small portion of harmonic energy will remain that will decrease following the roll-off performance of the filter.

the framework outlined in this chapter for 2D AESAs. The transmit case will be shown; however, using the same methodology, receive operation effects can also be derived.

Consider the HPA shown in Figure 4.17. A single signal with an applied phase shift for electronic scan is input into the HPA. From Equation 4.6, the output of the HPA will consist of the fundamental (transmit signal with frequency ω_o), and *nth*-order harmonics. As discussed previously, harmonics greater than the third order are either filtered out or have a low enough amplitude to not be considered. Using the expression for a 1D linear AESA, the *AF* for the *nth*-order output of the amplifier can be written as

$$AF_n = \sum_{n=1}^{N} a_n e^{j\left(n\omega t + x_n\left(\frac{\omega}{c} sin\theta - \frac{\omega_o}{c} sin\theta_o\right)\right)}, \ n = 1, 2, \ldots \tag{4.18}$$

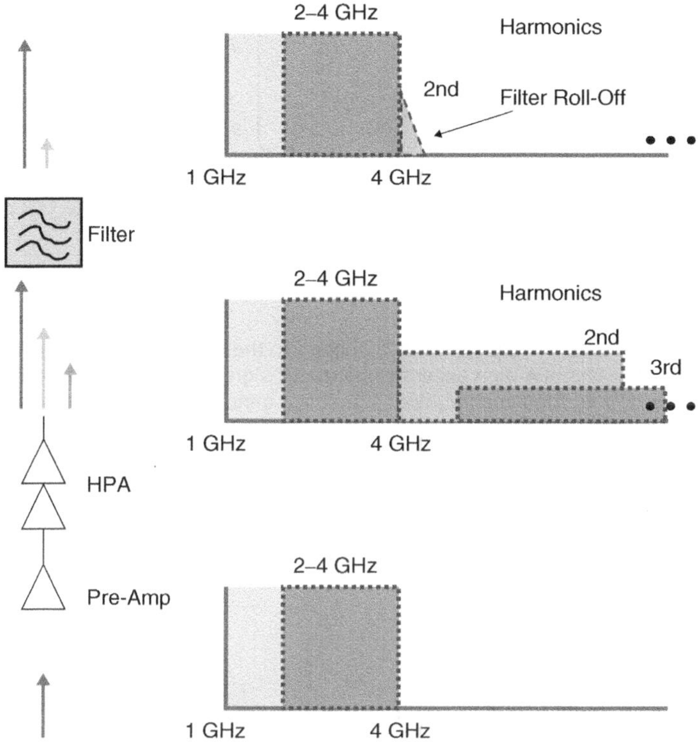

Figure 4.16 For multioctave AESAs, in-band signals can generate both in-band and out-of-band nonlinearities. The spurious products that are out of band will be filtered, but the spurious products will not be filtered and remain in-band.

The *nth*-order *AF* will have a maximum value when $\theta = sin^{-1}\left(\frac{\omega_o sin\theta_o}{\omega}\right)$, which when $\omega = \omega_0$ is θ_0. The fundamental, second-, third-, and higher-order harmonic *AFs* will have a maximum at the same scan angle (Hemmi, 2002) with narrower beamwidths due to their higher frequency.

For AESAs transmitting simultaneous signals, as shown in Figure 4.18, the same approach is used to calculate the angles where the nonlinear AFs have a maximum. This is outlined in depth in (Hemmi, 2002), and a summary from (Hemmi, 2002) is shown in Table 4.3.

4.2.5 Thermal Implications Due to Output Match

The active match of the array elements was covered in Chapter 3. It was shown that the active match changes with frequency and scan angle. This means that the output of the TRM is presented with a varying impedance, which is challenging

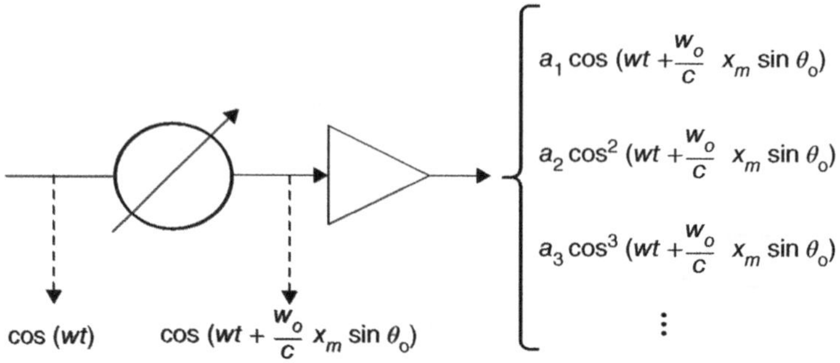

Figure 4.17 On transmit, a phase shift is applied to the transmitted signal prior to being input to the HPA. The HPA then generates harmonic signals that can coherently combine similar to the desired fundamental input signal and generate nonlinear harmonic beams.

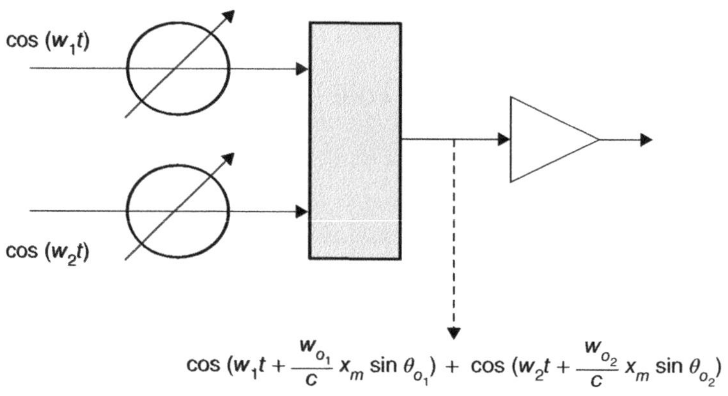

Figure 4.18 For two-tone input on transmit, two phase shifted signals are combined and then distributed to the HPA. These two phase shifted signals will generate harmonics and intermods that can then be radiated as beams from the AESA.

for AESAs since they are required to support both wideband and wide scan operation. From microwave circuit theory, a standing wave is produced when two RF circuits with different impedances are connected (Pozar, 2012). In Chapter 3, it was shown that the reflected power can be represented as $|\Gamma|^2$. While in transmit, the TRM sees both the transmitted power and the reflected power, which leads to an increased heat load for values of $|\Gamma|^2$ that are large. The power the TRM has to handle for thermal cooling is then $1 + |\Gamma|^2$. As an example, consider a TRM that has to transmit 100 W. For $|\Gamma| = 0.5$, 25 W (100 W $\cdot(0.5^2)$ of power would be reflected back at the TRM, and the total power the TRM would need to handle is 125 W. If for this example the AESA had 100 elements, that means an additional 2.5 kW would

Table 4.3 The table shows the locations of harmonic and intermod beams generated by the HPA on transmit due to the nonlinearities generated by the HPA

Description	Frequency	Beam Angle
Fundamental #1 (2nd/3rd harmonics)	$f_1(2f_1, 3f_1)$	θ_1
Fundamental #2 (2nd/3rd harmonics)	$f_2(2f_2, 3f_2)$	θ_2
Third-order difference	$2f_1 - f_2$	$\theta = \sin^{-1}\left\{\frac{2f_1 \sin\theta_1 - f_2 \sin\theta_2}{2f_1 - f_2}\right\}$
	$2f_2 - f_1$	$\theta = \sin^{-1}\left\{\frac{2f_2 \sin\theta_2 - f_1 \sin\theta_1}{2f_2 - f_1}\right\}$
Difference	$f_1 - f_2$	$\theta = \sin^{-1}\left\{\frac{f_1 \sin\theta_1 - f_2 \sin\theta_2}{f_1 - f_2}\right\}$
Sum	$f_1 + f_2$	$\theta = \sin^{-1}\left\{\frac{f_1 \sin\theta_1 + f_2 \sin\theta_2}{f_1 + f_2}\right\}$
Third-order sum	$2f_1 + f_2$	$\theta = \sin^{-1}\left\{\frac{2f_1 \sin\theta_1 + f_2 \sin\theta_2}{2f_1 - f_2}\right\}$
	$2f_2 + f_1$	$\theta = \sin^{-1}\left\{\frac{2f_2 \sin\theta_2 + f_1 \sin\theta_1}{2f_2 - f_1}\right\}$

need to be dissipated. This can drive the thermal design of the AESA if the reflected power is not minimized as much as possible over the required bandwidth and scan.

4.3 Receive Operation

In receive operation the LNA becomes the focus as it drives the receive performance of the TRM. The LNA is the first amplifier in the AESA RF chain and is one of the strongest contributing factors to noise figure and linearity. It is desired that the LNA have a gain ≥ 20 dB and low[1] NF ($NF = 10log_10(F)$). In any RF chain, the cascaded noise factor can be represented as (Pettai, 1984), (Pozar, 2012):

$$F_{cascaded} = F_1 + \frac{(F_2 - 1)}{G_1} + \frac{(F_3 - 1)}{G_1 \cdot G_2} + \cdots \tag{4.19}$$

In Equation 4.19, F_1 represents the noise factor of the LNA, and G_1 represent the gain of the LNA. For large values of G_1, $F_{cascaded} \approx F_1$. This is why the gain of the LNA is desired to be large with a small NF. As discussed in Chapter 1, the AESA NF directly affects the sensitivity ($\frac{G}{T}$) of the AESA system.

1 Low is relative depending on the frequency band and operational bandwidth. For limited bandwidth applications, ≈ 1 dB of NF noise figure for the LNA can be obtained, but for multioctave applications the NF is larger.

In addition to establishing the sensitivity performance, the TRM also introduces nonlinearities into the AESA on receive due to the LNA. The linearity principles discussed earlier for transmit operation are directly applicable for the TRM in receive. The difference is that the LNA instead of the HPA is the primary contributor. For wideband systems, a filter bank is typically placed after the LNA as shown in Figure 4.7. This enables harmonics and intermods to be filtered before the output of the AESA on receive and minimizes nonlinear signals presented to the RF conversion electronics (downconverters/upconverters) and the receiver. One of the primary differences on receive relative to the input power to the LNA is that the input power is composed of RF energy from the environment. This can be in the form of friendly or hostile emitters, and the frequency content can be in-band or out of band relative to the AESA operational bandwidth. This drives the LNA to have exceptional linearity performance in terms of its intercept points. Typically there are additional amplifier stages post LNA that can amplify the nonlinear signals generated by the LNA, which can be detrimental to receiver performance if they are not accounted for in the design.

Similarly, environmental interference on receive also impacts P_{1dB}. Very large signals from the environment that are incident on the AESA can saturate the TRM's LNA and also damage it. A receiver protector (see Figure 4.5) is used to account for damaging power levels, but for saturating power levels, the LNA is on its own. Therefore, analysis is done at the system level to determine the max environmental power that the LNA will be presented with, and this is then flowed down as a requirement for the P_{1dB} of the TRM on receive. The same process is followed for any amplifier stages post LNA as well. If any of the devices are saturated, the receiver is then presented with the maximum signal power that it can handle and reduces the overall system dynamic range.

4.4 Reliability[2]

At first glance, one may logically assume that the TRM is the biggest driver of the reliability of a system due to the large number of TRMs in an AESA. Counterintuitively, though, TRMs improve the reliability of the system and provide what is typically referred to as graceful degradation. The TRMs are not a source of single point of failure for an AESA[3]. The distributed nature of the TRM electronics enables minimal impact to system performance for P_{TX}, G, and SNR in the RRE. As discussed in Chapter 1, this was one of the most significant advantages in the maturation from MSAs to AESAs. In an MSA, there is a single HPA and LNA, and if they

2 This section is based on a white paper written by Mr. Bill Hopwood.
3 The same can be said for the array elements and also the beamformer; however, both of these are passive and already are characterized by high reliability.

fail, the system is no longer functional. For an AESA, a percentage of the TRMs could fail, and the AESA would remain operational with acceptable performance. This type of graceful degradation only applies to an AESA with a large number of elements. As an example, an AESA with 1000 elements will exhibit graceful degradation, but an AESA with 100 elements may not. For the remainder of this section, the analysis focuses only on AESAs with a large number of elements. Because large is not quantifiable, graceful degradation will also depend on the operational bandwidth; therefore, *large* must be evaluated on a case-by-case basis.

Different missions that use AESA systems also require different levels of reliability. For mission applications where the system has a low operational duty cycle, when the system is not in use, it is available for repair. In these situations, MTBF is applicable and is optimized to provide availability over many operational duty cycles. For applications where the the operational duty cycle is high and the system is not available for repair, MTBF is not very relevant. Instead, redundancy must be built into the system to ensure the probability of element failures is extremely low. In this situation, the TRM topology described in Figure 4.7 has great benefit. A system that has multiple simultaneous beams can revert to a single beam as a means of redundancy in case the other beam path(s) fail.

This section will describe a methodology for calculating the probability of element failures for an AESA and also the mean time between failures (MTBF). For an AESA system, system level reliability analysis is always performed; however, in this chapter the TRM will be the focus of the reliability discussion. Both element failures and MTBF are extremely important for characterizing the system. If enough elements failed, the AESA would not perform, and this would affect the availability of the system for operation. In systems that provide important intelligence information or protect the lives of the warfighter, the design must be robust from a reliability perspective. Additionally, failed elements can result in higher SLLs, which can degrade SL clutter reduction and increase the ability to be jammed by intentional/unintentional SL interference. Finally, as discussed in Chapter 2, failed elements will result in reduced P_{TX} and G.

4.4.1 Probability of Failed Elements

The binomial density function can be used to calculate the probability that a certain number of elements in the system will fail (Brown, 2012). This method assumes that the TRM is either fully operational or completely nonfunctional. For each element, the probability that the element is fully functioning, P, and the probability that the element is not working, $(1 - P)$, are assigned probabilities that sum to one. As an example, a four-element AESA is illustrated in Figure 4.19. In this example, the binomial method has been used to calculate the probability that four or fewer TRMs will fail in the example AESA shown in Figure 4.19.

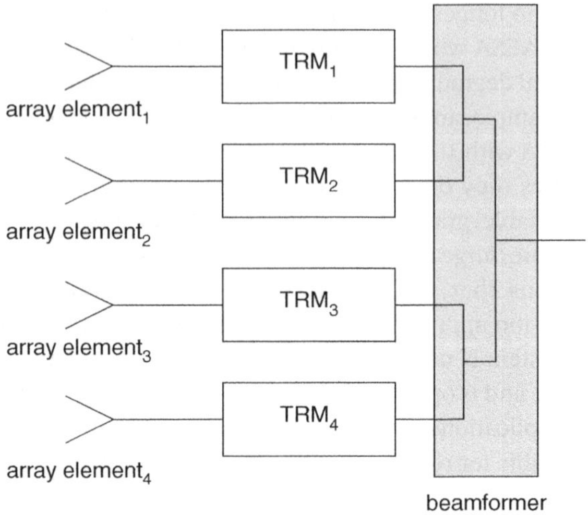

Figure 4.19 Four element AESA example used to calculate the probability of element failure.

A truth table showing all of the possible states for the AESA is shown in Table 4.4. The probability that all four modules are working or that there are zero failures is 0.6561, while the probability that all four modules are not working is 0.0001. The cumulative probability for each state is also calculated. The number of calculations required to determine the cumulative probability for a specific failure state increases as the number of elements is increased making the brute-force method very inefficient for large arrays. The binomial probability mass function can be used in these cases. The binomial probability mass function, $b(i, N, P)$, calculates the probability of exactly i successes in N trials, where P is the probability of each success. It assumes that the trials are independent. The binomial probability mass function is expressed as

$$b(i, N, P) = \frac{N!}{i!(N-i)!} P^i (1-P)^{N-i}. \tag{4.20}$$

The probability of F or fewer successes is then

$$P(\#\ of\ successes \leq F) = \sum_{i=0}^{F} b(i, N, P), \tag{4.21}$$

which is the sum of the individual probabilities. The binomial probability mass function is used to calculate the probabilities from the previous four-element AESA example. The probability that all four of the modules are working is

$$b(4, 4, 0.9) = \frac{4!0.9^4(1-0.9)^{(4-4)}}{4!(4-4)!} = 0.9^4 = 0.6561 \tag{4.22}$$

Table 4.4 Truth table and probability of element failures tabulated for the four-element AESA in Figure 4.19

N = Number of Elements, W = Working Elements, F = Failed Elements			
2^N States	**Element States**	**Failure Probability**	**F**
1	WWWW	.9 .9 .9 .9 = 0.6561	0
2	WWWF	.9 .9 .9 .1 = 0.0729	1
3	WWFW	.9 .9 .1 .9 = 0.0729	1
4	WWFF	.9 .9 .1 .1 = 0.0081	2
5	WFWW	.9 .1 .9 .9 = 0.0729	1
6	WFWF	.9. 1. 9 .1 = 0.0081	2
7	WFFW	.9. 1. 1 .9 = 0.0081	2
8	WFFF	.9. 1. 1 .1 = 0.0009	3
9	FWWW	.1. 9. 9 .9 = 0.0729	1
10	FWWF	.1 .9. 9 .1 = 0.0081	2
11	FWFW	.1 .9 .1 .9 = 0.0081	2
12	FWFF	.1 .9 .1 .1 = 0.0009	3
13	FFWW	.1 .1 .9 .9 = 0.0081	2
14	FFWF	.1 .1 .9 .1 = 0.0009	3
15	FFFW	.1 .1 .1 .9 = 0.0009	3
16	FFFF	.1 .1 .1 .1 = 0.0001	4

Prob of 4 or Fewer Failures = 0.6561 + 0.2961 + 0.0486 + 0.0036 + 0.0001 = 1			
States with F Failures	**Number of States**	**Prob of F Failures**	**Cum Prob of F Failures**
F=0	1	0.6561	0.6561
F=1	4	0.0729	0.2916
F=2	6	0.0081	0.0486
F=3	4	0.0009	0.0036
F=4	1	0.0001	0.0001

Similarly, the probability that three, two, one, or zero of the four modules are working is

$$b(3, 4, 0.9) = \frac{4!0.9^3(1-0.9)^{4-3}}{3!(4-3)!} = \frac{4 \times 3 \times 2 \times 1}{3 \times 2 \times 1} 0.9^3 0.1 = 0.2916 \qquad (4.23)$$

$$b(2,4,0.9) = \frac{4!0.9^2(1-0.9)^{4-2}}{2!(4-2)!} = \frac{4\times3\times2\times1}{2\times1\times2\times1}0.9^2 0.1^2 = 0.0486 \tag{4.24}$$

$$b(1,4,0.9) = \frac{4!0.9^1(1-0.9)^{4-1}}{4!(4-1)!} = \frac{4\times3\times2\times1}{1\times3\times2\times1}0.9^1 0.1^3 = 0.0036 \tag{4.25}$$

$$b(0,4,0.9) = \frac{4!0.9^0(1-0.9)^{4-0}}{4!(4-0)!} = \frac{4\times3\times2\times1}{1\times4\times3\times2\times1}0.9^0 0.1^4 = 0.0001 \tag{4.26}$$

$$0.6561 + 0.2916 + 0.0486 + 0.0036 + 0.0001 = 1 \tag{4.27}$$

The sum of the five probabilities is equal to one as previously shown in Table 4.4. The binomial distribution function is an efficient means of calculating the probability of the number of working modules for a large AESA.

4.4.2 MTBF

MTBF is the average time between failures of a system. For an AESA this can be quantified with the following expression (Brown, 2012):

$$MTBF = \frac{\textit{total operating time with a large element population}}{\textit{total failures during that time}}. \tag{4.28}$$

MTBF provides a way to quantify the predicted amount of time between failures. As an example for computing MTBF, consider an AESA with 200 TRMs that have operated for 10,000 hours with 5 total failures. Using Equation (4.28), the MTBF is calculated as

$$MTBF_{AESA} = \frac{200 \times 10{,}000}{5} = 400{,}000 \text{ hours}. \tag{4.29}$$

In designing an AESA, the failure rate of the TRM is used to compute the MTBF giving an alternative formulation that relates the failure rate of the TRM to the MTBF for the AESA ($MTBF_{AESA}$). This approach is described next.

Typically, there is a system level requirement for MTBF that is flowed down to the TRM. This then drives the TRM design in terms of failure rate. Because the TRMs are in a large distributed population in the AESA, a single failure does not fail the entire AESA or system. For improved robustness, the MTBF for a TRM can be improved by increasing the integration of MMICs, improving the manufacturing processes, or reducing the module operational temperature (Agrawal and Holzman, 1999). To determine the number of failures that are acceptable, an allowable peak SLL and/or gain reduction is used to determine the maximum failure percentage of elements. (Agrawal and Holzman, 1999) provides an example using SLLs for determining allowable fractions of element failures.

Taking the element failure rate into account, the $MTBF_{AESA}$ can be calculated using the reliability function, $R(t)$, which is shown in Equation 4.30 for a single element.

$$MTBF_E = \int_0^\infty R_E(t)dt, \tag{4.30}$$

where $MTBF_E$ is the MTBF of an element and $R_E(t)$ can be defined as

$$R_E(t) = e^{-\lambda_E t}, \tag{4.31}$$

where λ_E is the element failure rate. This leads to

$$MTBF_E = \int_0^{\infty} e^{-\lambda_E t} dt = \frac{1}{\lambda_E}. \tag{4.32}$$

Using the binomial distribution function, $MTBF_{AESA}$ can be defined as F or fewer failed elements using

$$R_{AESA}(t) = \sum_{i=0}^{F} b(i, N, R_E(t)), \tag{4.33}$$

where R_{AESA} is the AESA reliability function. The AESA MTBF is then

$$MTBF_{AESA} = \int_0^{\infty} R_{AESA}(t)dt \tag{4.34}$$

Using Equations 4.33 and 4.34 $MTBF_{AESA}$ can be expressed as

$$MTBF_{AESA} = \sum_{i=0}^{F} \int_0^{\infty} \frac{N!}{i!(N-i)!}(1 - e^{-\lambda_E t})_i e^{-\lambda_E t(N-i)} dt \tag{4.35}$$

Upon manipulation, this reduces to

$$MTBF_{AESA} = \frac{1}{\lambda_E} \sum_{i=0}^{F} \frac{1}{N-i} \approx \frac{1}{\lambda_E} \frac{F}{N} = MTBF_E \frac{F}{N}. \tag{4.36}$$

Thus, $MTBF_E$ scales $MTBF_{AESA}$ by the fraction of failed elements. As an example, if $MTBF_E$ is 1E6 hours then λ_E is $1E-6$, and if 5% failures are allowable ($F/N = 0.05$), the $MTBF_{AESA}$ is $0.05 \times 1E6 = 50{,}000$ hours.

The previous derivation has assumed that an element failure is the same as a TRM failure. In actual systems, the failure rate also must take into account additional components that support the TRM such as power supplies and control modules. The MTBF expression used in Equation 4.36 can also be applied to the additional TRM supporting components as well and is shown in (Agrawal and Holzman, 1999).

In addition to reliability, availability is also an important parameter and can be expressed as (Agrawal and Holzman, 1999):

$$availability = \frac{MTBF_{AESA}}{MTBF_{AESA} + MTTR_{AESA}}, \tag{4.37}$$

where $MTTR_{AESA}$ is the mean time to repair or replace failed components in the AESA (Agrawal and Holzman, 1999). The availability is maximized by increasing $MTBF_{AESA}$ and/or decreasing $MTTR_{AESA}$.

References

Agrawal, A., Clark, R., and Komiak, J. "T/r module architecture tradeoffs for phased array antennas." *IEEE MTT-S International Microwave Symposium Digest*, pp. 995–998, 1999.

Agrawal, A.K., and Holzman, E.L. "Active phased array design for high reliability." *IEEE Transactions on Aerospace and Electronic Systems*, pp. 1204–1211, 1999.

Brown A.D. *Electronically Scanned Arrays: MATLAB® Modeling and Simulation*. CRC Press, 2012.

Chun, W., Kexin, L., Mingliang, H., and Jing, D. "Comments on 'pattern characteristics of harmonic and intermodulation products in broad-band active transmit arrays.'" 2016 11th International Symposium on Antennas, Propagation and EM Theory, pp. 31–34, 2016.

Hemmi, C. "Pattern characteristics of harmonic and intermodulation products in broadband active transmit arrays." *IEEE Transactions on Antennas and Propagation*, pp. 858–865, 2002.

Holter, H., and Steyskal, H. "On the size requirement for finite phased-array models." *IEEE Transactions on Antennas and Propagation*, pp. 836–840, 2002.

Loyka, S. "Comments on 'Pattern characteristics of harmonic and intermodulation products in broad-band active transmit arrays'." *IEEE Transactions on Antennas and Propagation*, p. 1683, 2003.

McQuiddy Jr., D.N, Gassner, R.L., Hull, P., Mason, J.S., and Bedinger, J.M. "Transmit/receive module technology for x-band active array radar." *Proceedings of the IEEE*, pp. 308–341, 1991.

Pettai, R. *Noise in Receiving Systems*. John Wiley & Sons, 1984.

Pozar, D.M.. *Microwave Engineering*. John Wiley & Sons, Inc., 2012.

Sandrin, W. "Spatial distribution of intermodulation products in active phased array antennas." *IEEE Transactions on Antennas and Propagation*, pp. 864–868, 1973.

Walker, J.L.B. *Handbook of RF and Microwave Power Amplifiers*. Cambridge University Press, 2012.

5

Beamformers

Key Concepts

- Tile vs. Brick Architectures
- Corporate vs. Noncorporate
- Lossless Beamformer
- Amplitude Weighting
- Beam Spoiling
- Monopulse Beamforming

5.1 Introduction

The beamformer in an AESA performs the function of distributing RF energy on transmit and combining RF energy on receive. It is sometimes referred to interchangeably as a manifold because of the distribution function it performs. Figure 5.1 shows the location of the beamformer in the AESA block diagram.

In transmit operation, the beamformer takes the transmit power via a modulated voltage waveform that is generated at the system exciter. This voltage is distributed to each of the elements by the beamformer, amplified by the TRMs with an appropriately applied phase shift, and then transmitted spatially by the array elements. This is shown in Figure 5.2. The transmitted beam is formed in the far-field of the array at the commanded scan angle θ_o.

In receive operation, instead of distributing voltages, the beamformer combines the received voltages from the array elements with the correct phase shift applied by the TRM. This is shown in Figure 5.3. The output of the beamformer is then sent to the receiver.

Active Electronically Scanned Arrays: Fundamentals and Applications, First Edition. Arik D. Brown.

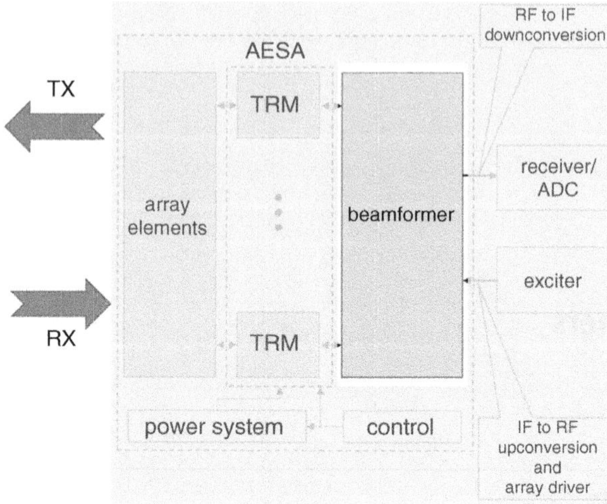

Figure 5.1 The beamformer in the AESA distributes RF energy on transmit and combines RF energy on receive.

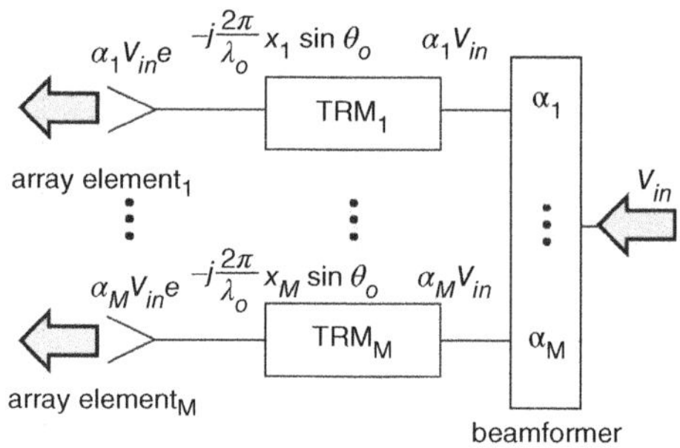

Figure 5.2 In transmit operation, the AESA beamformer distributes the transmit waveform voltage to each of the array elements.

5.1.1 Tile and Brick Architectures

The beamformer is typically constructed in a printed circuit board style using stripline or microstrip circuitry for RF voltage distribution. Other applicable transmission media for the beamformer are waveguides and coaxial lines [Mailloux, 1993]. For modern AESAs, microstrip/stripline is typically used because they are less bulky and more easily modeled and manufactured for large AESAs.

An architectural constraint to the beamformer is whether the AESA is constructed using a brick or tile approach. This refers to the way in which the AESA is assembled. In a brick architecture the TRMs are discrete hardware assemblies

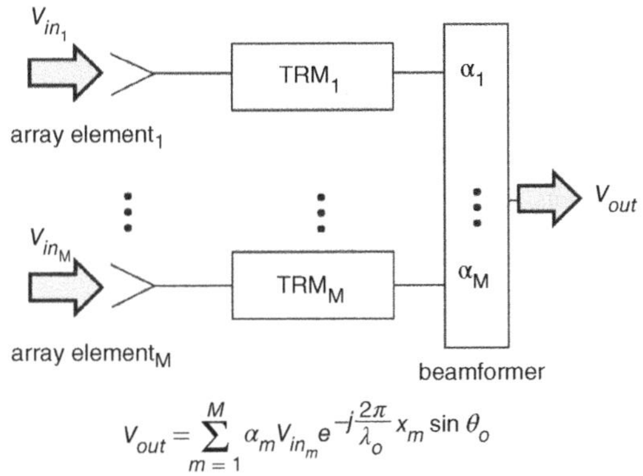

Figure 5.3 In receive operation, the AESA beamformer combines the signals received at each array element.

containing one or multiple transmit and receive channels. They are mounted perpendicular to the array face and have a larger depth than a tile approach [Mailloux, 1992]. Figure 5.4 shows a brick architecture. The beamformer would plug into the perpendicular TRMs.

In applications where the available space is not constrained, such as in a ship, a brick architecture is attractive for its thermal cooling advantage. The depth provides more opportunity for thermal heat dissipation compared to a tile architecture. An additional advantage of the the brick architecture is its compatibility with naturally wideband elements such as notches (discretized Vivaldi/flared element) [Mailloux, 1992]. However, from a weight perspective the brick-style architecture will be at a disadvantage for compact applications with challenging SWaP requirements. For brick architectures the beamformer is typically not integrated with the array element and TRM assemblies.

Tile architectures can be described by what their name implies. The array elements, TRMs, and beamformer are integrated into two-dimensional planar layers that resemble tiles. This is illustrated in Figure 5.5. The RF signals for tile AESAs are distributed by a beamformer that is parallel to the array face. Tile-constructed AESAs are typically lower weight and extremely compact, making them suitable for applications with challenging size and weight requirements. They are also inexpensive to produce and compatible with automatic means of fabrication [Mailloux, 1992]. Tile architectures are also favorable for subarray (SA) applications. Making an SA building block that is the size of a tile leads to a highly producible design that can minimize cost. The SA tiles are manufacturing building blocks that support scalability for applications requiring similar capability but have different ERP or sensitivity requirements. By simply scaling the number of SA tiles, the same product baseline can be

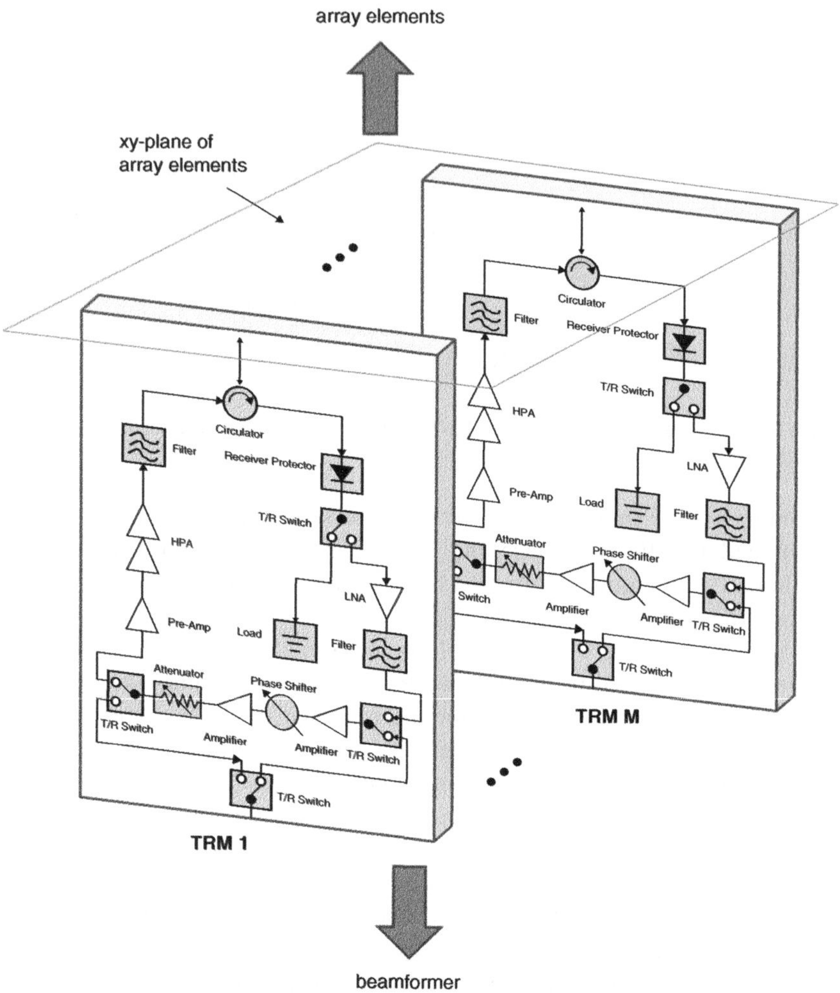

Figure 5.4 In an AESA brick architecture, the TRM assemblies are mounted perpendicular to the plane of the array elements. The beamformer is connected to the side of the TRMs opposite the array elements.

used for multiple applications. Additionally, a SA tile that supports transmit and receive operation can be used for receive-only simply by removing the transmit circuitry. This can substantially decrease the non recurring cost for an AESA. The required capability is available with the tile SA design and requires minor modifications.

Instead of using RF circuitry for the beamformer to distribute and combine RF energy, a space-fed approach can also be used. This approach has the benefit of

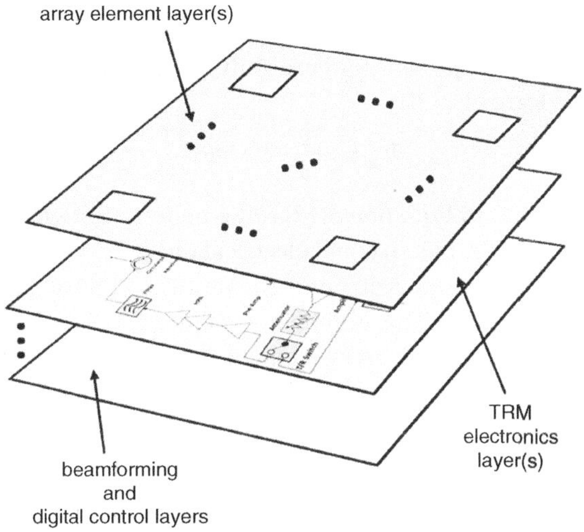

Figure 5.5 In AESA tile architectures, the beamformer is part of an integrated multi-layer stackup that also includes digital control, planar TRM circuitry, and planar array elements.

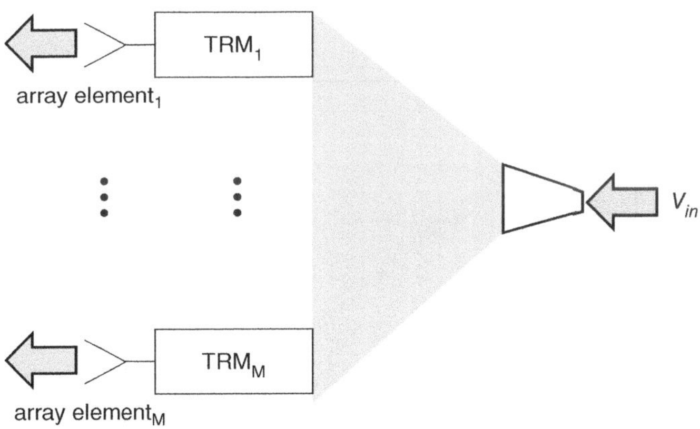

Figure 5.6 Space-fed beamformers use free space for signal distribution without the need for RF microwave beamformer circuitry.

less loss since the distribution transmission medium is free space as shown in Figure 5.6. Space feeds, however, require a significant increase in volume and are not very amenable to compact and/or low profile AESAs [Mailloux, 1992]. This chapter will primarily focus on constrained beamformers that employ RF circuitry for distribution of RF energy in the AESA.

5.1.2 Corporate and Noncorporate Beamforming

The TRMs in a 1D AESA provide a phase shift, Φ_m, to each element on transmit and receive in the form of:

$$\Phi_m = -\frac{2\pi}{\lambda_o} x_m sin\theta_o. \tag{5.1}$$

The applied phase shift in Equation 5.1 has the underlying assumption that each channel path in the AESA has an equal electrical line length. This produces a progressive phase shift across the array and enables the AESA to scan the beam to the commanded scan angle. This is shown in Figure 5.7. A beamformer that has equal electrical line lengths for every element is called a *corporate* beamformer. This style of beamformer is composed of multiple power splitters. When the total number of elements, M, in the AESA is a power of 2, the total number of power dividers per element channel can be calculated as

$$\begin{aligned} \#\ of\ power\ dividers &= log_2(M) \\ &= \frac{log_{10}(M)}{log_{10}(2)} \end{aligned} \tag{5.2}$$

The number of power dividers drives the loss in the beamformer. As an example, an AESA with a 1024 (32 × 32) element beamformer would have 10 power dividers

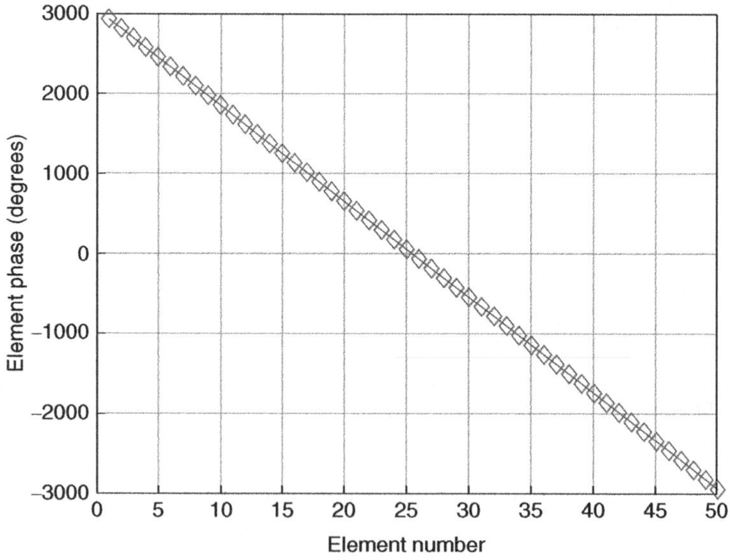

Figure 5.7 The phase at each element for a 50-element 1D AESA. A progressive phase shift is applied to each element for electronic scanning.

The outputs of the beamformer are represented by v_m, where $m = 1, 2, ..., M$. α_m are the beamformer weights for each element path. On transmit, the beamformer is an $M + 1$ port power divider with one input and M outputs. Using conservation of energy, for a lossless beamformer the input power must equal the output power [Balanis, 1982]. The input power for the beamformer in Figure 5.9 is:

$$P_{in} = |v_o|^2. \tag{5.4}$$

The corresponding output power in Figure 5.9 is:

$$\begin{aligned} P_{out} &= \sum_{m=1}^{M} |\alpha_m v_o|^2 \\ &= |\alpha_m v_o|^2 \sum_{m=1}^{M} 1 \\ &= \alpha_m^2 |v_o|^2 M \\ &= \alpha_m^2 P_{in} M. \end{aligned} \tag{5.5}$$

Conservation of energy dictates that for the lossless beamformer, $P_{in} = P_{out}$, which means that in Equation 5.5, $\alpha_m^2 M = 1$, so $\alpha_m = \frac{1}{\sqrt{M}}$. The output element voltage can then be expressed as:

$$v_m = \frac{1}{\sqrt{M}} v_o. \tag{5.6}$$

5.2.2 Receive

On receive, a similar formulation can be derived that expresses the combined output of the M inputs of the beamformer. A receive beamformer is shown in Figure 5.10 that is effectively an M:1 combiner. The voltage at each input port is

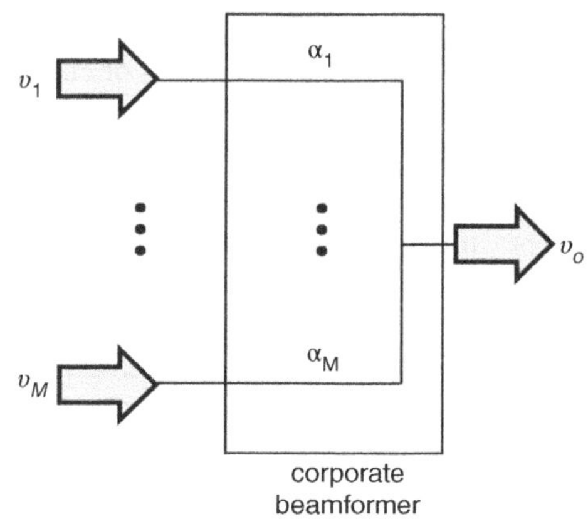

Figure 5.10 The figure illustrates an M:1 corporate receive beamformer. The voltage signals from all M elements are added together coherently in the beamformer.

υ_m, and the output voltage is υ_{out} and is expressed as:

$$\upsilon_{out} = \sum_{m=1}^{M} \alpha_m \upsilon_m. \tag{5.7}$$

For the lossless beamformer, conservation of energy requires the total input power of the beamformer to be equal to the output power. The input power can be formulated as:

$$P_{in} = \sum_{m=1}^{M} |\upsilon_m|^2, \tag{5.8}$$

and the output power can be represented as

$$P_{out} = \sum_{m=1}^{M} |\alpha_m \upsilon_m|^2 \tag{5.9}$$

It is assumed that $|\upsilon_m|$ is constant for $m = 1, 2, ..., M$, and $|\upsilon_m| = |\upsilon_{in}|$ for $m = 1, 2, ..., M$. Since the TRM applies a phase shift for each element (phase does not change the voltage magnitude), this is a valid assumption. This will be revisited when distributed weighting (amplitude taper for low SLLs is distributed between the TRM attenuators and the beamformer) is covered later. This assumption simplifies Equation 5.8 to

$$P_{in} = M|\upsilon_{in}|^2. \tag{5.10}$$

For the output power, it is assumed that the input voltages from the TRMs are phased appropriately such that the sum of the voltages adds to M. This is exactly the situation described in Chapter 1 where the sum of the AF is equal to the number of elements in the AESA at the commanded scan angle. Additionally, uniform distribution is assumed such that $\alpha_m = \alpha_o$ for $m = 1, 2, ..., M$. With these assumptions, Equation 5.9, can be simplified to

$$P_{out} = \alpha_o^2 M^2 |\upsilon_m|^2. \tag{5.11}$$

Since P_{out} must be equal to P_{in} due to conservation of energy, the result is that, similar to the transmit beamformer, $\alpha_m = \alpha_o = \frac{1}{\sqrt{M}}$, and $P_{out} = M|\upsilon_m|^2$, which is equivalent to Equation 5.8 for P_{in}. Intuitively, this makes sense. The beamformer provides a signal gain of M for the AESA.

An important concept can be described using the equation for P_{out}. The SNR at the beamformer output can be expressed as $N_{out} = k, T, B$ and F. Because noise does not coherently add, the noise at the output of the beamformer is simply the input noise, $N_{out} = kTB$ scaled by F, which in this case is the loss of the beamformer. The output SNR of the beamformer can be written as,

$$\begin{aligned} SNR_{output} &= \frac{M|\upsilon_m|^2}{kTBF} \\ &= \frac{M|\upsilon_m|^2}{kTB}, \; F = 1 \; \textit{for a lossless beamformer}. \end{aligned} \tag{5.12}$$

Equation 5.12 shows that the output SNR of a beamformer is increased by the number of inputs to the beamformer and degraded by the loss in the beamformer. The term *inputs* is used instead of elements since for a SA architecture, the SA beamformer will be a subset of the number of elements in the AESA.

An additional concept affecting AESA performance can be shown using Equation 5.12. When there are failed elements, P_{out} must be modified using Equation 5.11 as shown in Equation 5.13. In Equation 5.13, ϵ is the fraction of failed elements.

$$\begin{aligned} P_{out} &= \alpha_o^2[M(1-\epsilon)]^2|v_m|^2 \\ &= \frac{1}{M}[M(1-\epsilon)]^2|v_m|^2 \\ &= M(1-\epsilon)^2|v_m|^2. \end{aligned} \tag{5.13}$$

In the case of failed elements, the output SNR of the beamformer can then be expressed as

$$SNR_{output} = \frac{M(1-\epsilon)^2|v_m|^2}{kTBF}. \tag{5.14}$$

This is very important because it shows that the degradation to SNR is a function of the fraction of failed elements squared. As an example, if half of the elements in the AESA failed, $\epsilon = 0.5$, then the SNR would be degraded by 6 dB. This supports the degradation on receive shown in Appendix E. This is readily seen from Equation 5.14 where $M(1-\epsilon)$ is equal to the number of failed elements.

5.3 Beamformer Weighting

Sidelobes provide an opportunity for unintentional or intentional interference to enter an AESA. Electronic countermeasure (ECM) jamming signals can enter the sidelobes using narrowband or wideband noise or with energy in the operational frequency range of the AESA. This can create false targets at the receiver or blind the receiver by raising the noise level above detectable signals. Radar clutter can also enter the AESA through the sidelobes and similarly to jamming signals, create false targets and/or blind the receiver. In order to counteract this, amplitude weighting can be applied to the array elements to reduce the sidelobe level.

Typically, when amplitude weighting is discussed, it is applied to the receive operation. This is because there is an associated loss with amplitude weighting that would affect the transmitted ERP. If the transmitted power is reduced, that places an added burden to increase the amount of prime power required to power the TRMs. For most systems, prime power is minimized as much as possible to reduce its thermal and cost impact of design accommodations. In standard operation, a uniform amplitude taper is applied on transmit to the AESA.

Amplitude weighting was discussed in Chapter 2. An example of different weightings is shown in Figure 5.11 using different Taylor distributions. Other weightings can be used, but as shown in [Taylor, 1955], Taylor weighting is the most efficient aperture distribution meaning it incurs the least loss in terms of signal gain. Figure 5.12 shows the trade-off that must be taken into consideration when using amplitude weighting. In Figure 5.12, a pattern with uniform distribution is shown in the top figure, and a pattern with 30 dB Taylor distribution is shown in the bottom figure. From these figures, two observations can be made. First, there is a beam broadening of the main beam when weighting is used. This translates to a loss in signal gain relative to SNR. Second, when the AESA is scanned, the 30 dB SL reduction is not maintained due to the loss in gain of the main beam when scanned. What this implies is that if a certain SL reduction is required across scan, then a heavier weighting must be applied so that the SL reduction is maintained at the larger scan angles. As an example, if a 30 dB SLL is required over all scan angles, then a weighting larger than 30 dB would need to be implemented. This trade-off between loss of gain and low SLLs is one that is well worth it to preserve AESA performance. This bears mentioning again: the Taylor distribution is used to keep the loss of gain as low as possible.

In Chapter 2, the loss due to amplitude weighting was referred to as taper loss. Building on the formulation for a lossless beamformer previously described, the

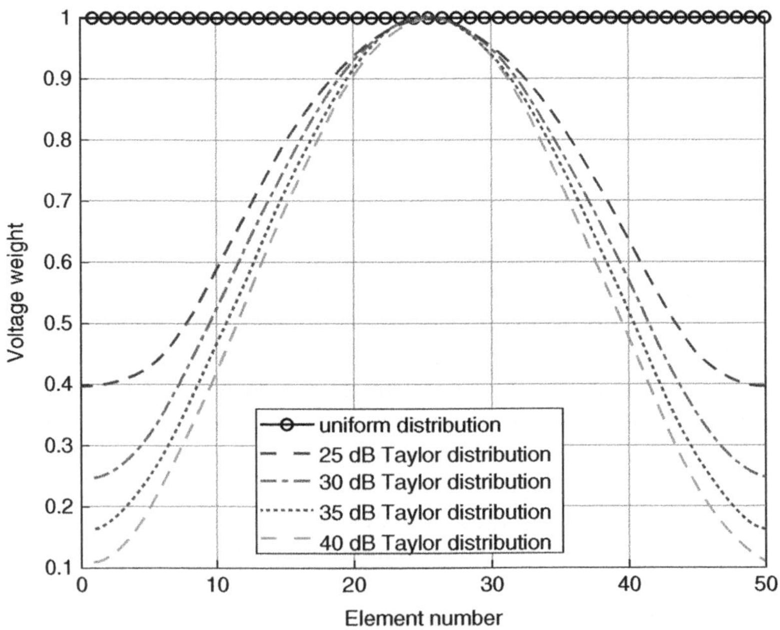

Figure 5.11 Various amplitude weightings compared to uniform weighting.

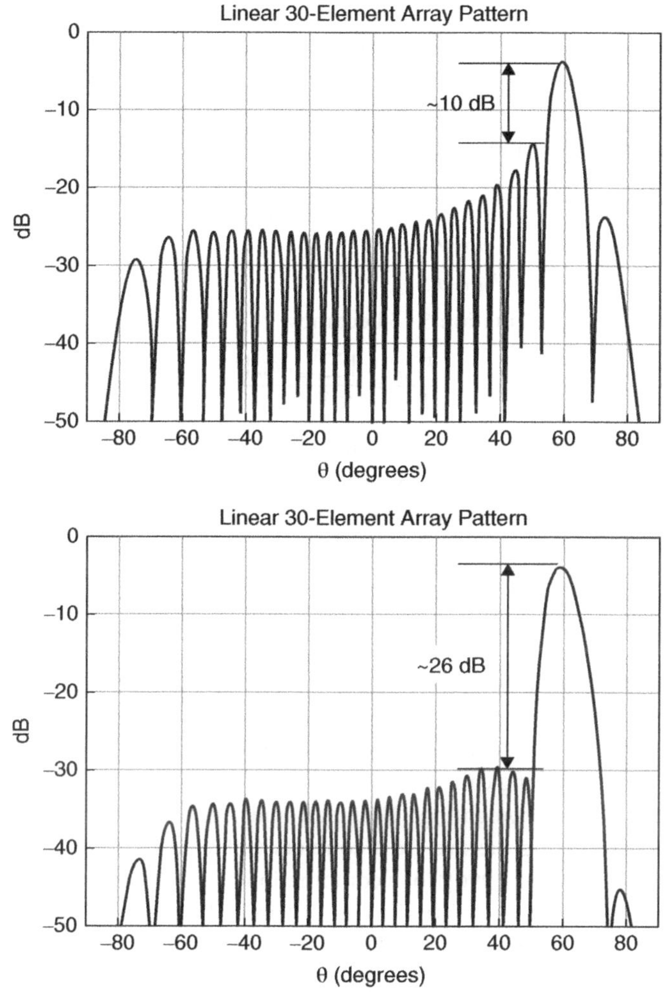

Figure 5.12 The main advantage of amplitude weighting is the reduction in SLLs. The bottom figure has a 30 dB Taylor distribution. When the AESA is scanned, the beam broadens in comparison to the uniform distribution (top figure). This is the trade-off when using amplitude weighting.

derivation of the taper loss expression can be shown. Using Equations 5.8 and 5.9, the loss in signal gain can be written as:

$$\frac{P_{out}}{P_{in}} = \frac{|\sum_{m=1}^{M} \alpha_m v_m|^2}{\sum_{m=1}^{M} |\alpha_m v_m|^2}, \tag{5.15}$$

where P_{in} has been modified to include the internal voltages of the beamformer for each element ($\alpha_m v_m$). Recall that the expression for the input voltage is:

$$v_m = e^{j\left(\frac{2\pi f}{c}\sin\theta - \frac{2\pi f_o}{c}\sin\theta_o\right)}. \tag{5.16}$$

At the tune frequency and commanded scan angle, the voltage v_m reduces to 1 as shown in Equation 5.17.

$$\begin{aligned} v_m|_{\theta=\theta_o, f=f_o} &= e^{\left(\frac{2\pi f_o}{c}\sin\theta_o - \frac{2\pi f_o}{c}\sin\theta_o\right)} \\ v_m|_{\theta=\theta_o, f=f_o} &= 1. \end{aligned} \tag{5.17}$$

Using Equation 5.17, Equation 5.15 becomes

$$\frac{P_{out}}{P_{in}} = \frac{|\sum_{m=1}^{M} \alpha_m|^2}{\sum_{m=1}^{M} |\alpha_m|^2} \tag{5.18}$$

and has a maximum value of M. Dividing by M, the expression in Equation 5.18 becomes the equation for taper loss as show in Equation 2.54 and is also referred to as taper efficiency [Mailloux, 1993].

$$TL = \frac{1}{M}\frac{P_{out}}{P_{in}} = \frac{1}{M} \cdot \frac{|\sum_{m=1}^{M} \alpha_m|^2}{\sum_{m=1}^{M} |\alpha_m|^2}. \tag{5.19}$$

5.4 Distributed Weighting

There are various approaches that can be used to employ amplitude weighting for an AESA. The weighting can be done completely in the TRMs using attenuators (referred to as active weighting [Holzman and Agrawal, 1996]), completely in the beamformer using passive 3 dB couplers (referred to as passive weighting [Holzman and Agrawal, 1996]), or a combination of the aforementioned methods. This last method will be referred to as distributed weighting since it refers to the amplitude weighting being distributed between both the TRMs and the beamformer.

As shown in [Holzman and Agrawal, 1996], applying the weighting with the TRMs only results in higher NF, reduced sensitivity $\frac{G}{T}$, and a lower second- and third-order order intercept point performance (reduced linearity). Applying the weighting completely in the beamformer provides the best sensitivity and linearity performance. In applications that require multiple levels of beamforming (subarrayed architectures) or where multiple tapers are required, it can be advantageous to use distributed beamforming, as shown in Figure 5.13. As an example, an AESA could use a nominal 25 dB amplitude weighting in the beamformer for normal operation and use the TRMs to apply additional weighting for a low SL mode of operation. Using the distributed approach in this example would eliminate the need for multiple beamformers and provide design flexibility.

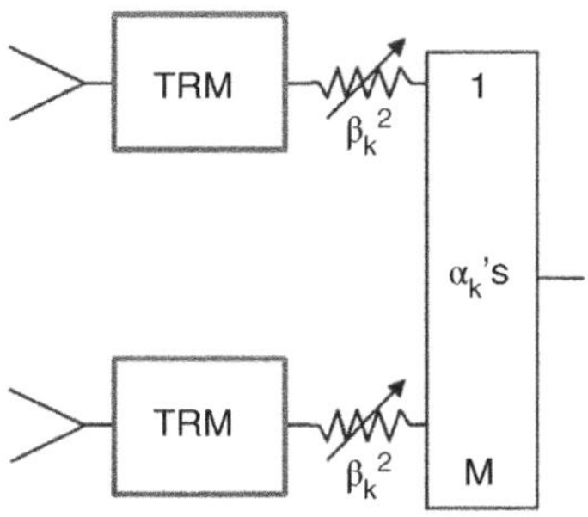

Figure 5.13 Distributed beamforming applies amplitude weighting in both the TRMs and the beamformer for an AESA.

5.5 Beam Spoiling

Thus far, only amplitude weighting has been discussed. Phase weighting can also be employed, but its primary use is not for SLL control. Instead, phase weighting enables beam spoiling, which is a way to increase the beamwidth of an AESA by using phase instead of amplitude weighting. This is done in applications such as radar and SIGINT, where the ability to increase the spatial scan rate of the FOV is of great benefit to the system. Figure 5.14 shows an AESA spoiled transmit beam and four simultaneous receive beams all normalized to their own peak value. The transmit beam has a spoil factor of ~ 2x meaning the 3 dB beamwidth is widened to a width that is twice that of the unspoiled beamwidth. The spoiled beam in Figure 5.14 is also applicable for receive operation.

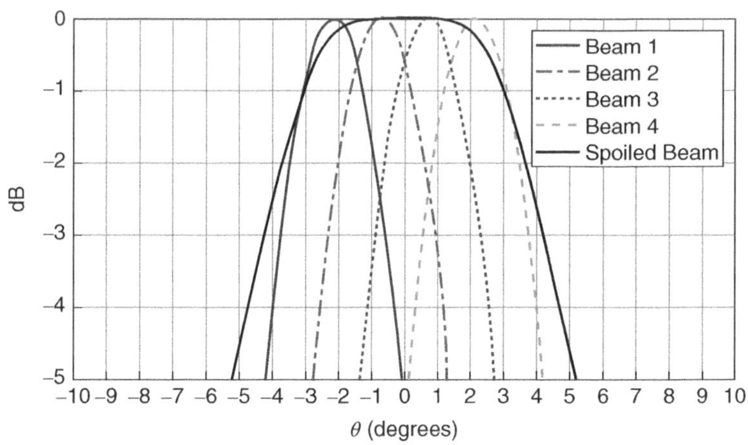

Figure 5.14 Spoiled transmit beam with a spoil factor of ~ 2x and four simultaneous receive beams. All the beams are normalized to their own peak values.

Various optimization approaches have been used for phase-only pattern synthesis with examples shown in [Brown et al., 2006], [Hu et al., 2006], and [Lin and Wang, 2000]. For beam spoiling, perhaps the most straightforward approach is a quadratic phase implementation. With this method, the phases for the weighting are generated by a quadratic expression with a scale factor. This section will describe how the phases are calculated and provide examples of different spoil factors. This methodology is also described in [Sayidmarie and Sultan, 2013].

A 1D M-element linear AESA will be considered. The same approach can be applied to a 2D AESA with no loss of generality. For each array element in the AESA, a phase index value is assigned in the following form:

$$\phi_m = \left(m - \frac{(M+1)}{2}\right)\left(\frac{2\psi\sqrt{\pi}}{M-1}\right), \quad m = 1, \ldots, M \tag{5.20}$$

The variable ψ in Equation 5.20 controls the spoil factor of the array beam. Using the expression for ϕ_m in Equation 5.20, the phases for the beam spoiling weights are calculated as

$$\Phi_m = \phi_m^2 \tag{5.21}$$

and the phase only weights as

$$\begin{aligned}\beta_{m_{phase\ only}} &= e^{j\Phi_m} \\ &= e^{j\phi_m^2} \\ &= e^{j\left[\left(m-\frac{(M+1)}{2}\right)\left(\frac{2\psi\sqrt{\pi}}{M-1}\right)\right]^2}.\end{aligned} \tag{5.22}$$

$\beta_{m_{phase\ only}}$ can be used to modify the expression for the 1D AESA pattern and is shown in Equation 5.23.

$$AF = \sum_{m=1}^{M} \beta_{m_{phase\ only}} e^{j\left(\frac{2\pi}{\lambda}x_m sin\theta - \frac{2\pi}{\lambda_o}x_m sin\theta_o\right)}. \tag{5.23}$$

Figure 5.15 shows the pattern of a 30 dB Taylor weighted unspoiled beam and a 30 dB Taylor weighted beam with a spoil factor of 2x. Both are normalized to their own peak values for illustration. The 1D array is composed of 40 array elements ($M = 40$).

The increased width of the spoiled beam does create a challenge for mitigating interference. An opportunity is provided for interfering signals to enter the AESA through the widened beamwidth. This can be mitigated with the appropriate selection of an auxiliary antenna for sidelobe blanking that includes the widened main beam region. Figure 5.16 shows a zoomed-in view of the 3 dB main beam regions for both the unspoiled and spoiled beams in Figure 5.15. Using phase-only weighting enables beamwidth flexibility. Figures 5.17 and 5.18 show the unspoiled and

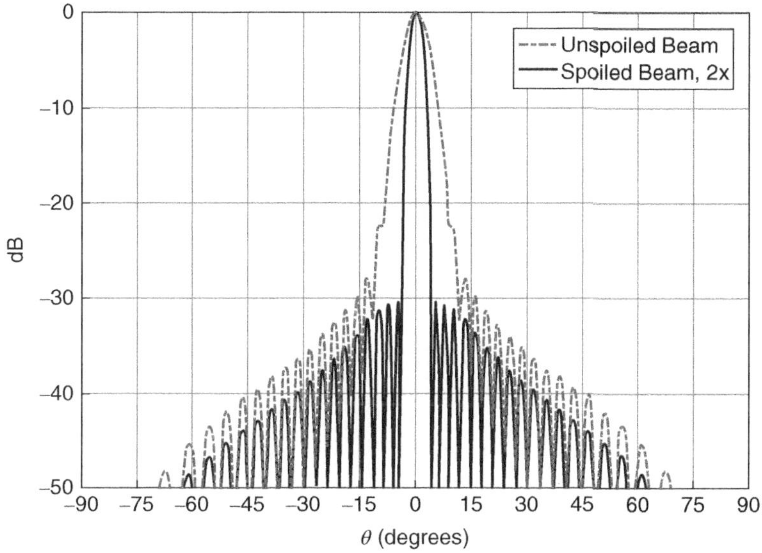

Figure 5.15 Spoiled beam generated using quadratic phase weighting for a 2x increase in the 3 dB beamwidth. Both patterns are normalized to their own peak values.

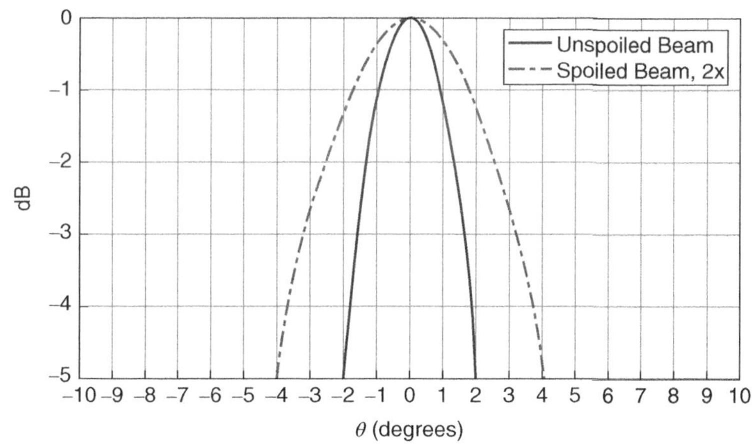

Figure 5.16 Zoomed-in view of the patterns shown in Figure 5.15. Both patterns are normalized to their own peak values.

spoiled patterns scanned to 60°. The weighting exhibits the same behavior for electronic scan of the AESA.

Figure 5.19 shows spoiled beams for the same 40-element AESA for 2x, 3x, and 4x beam broadening. The previous figures normalized the individual patterns to

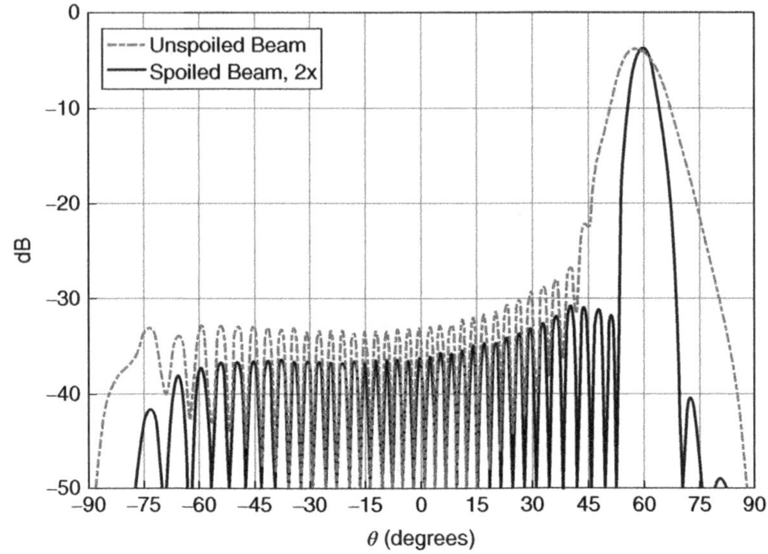

Figure 5.17 Spoiled beam scanned to 60°. The beamwidth increases due to scanning of the beam, as shown in Chapter 2, and the quadratic phase weighting. Both patterns are normalized to their own peak values.

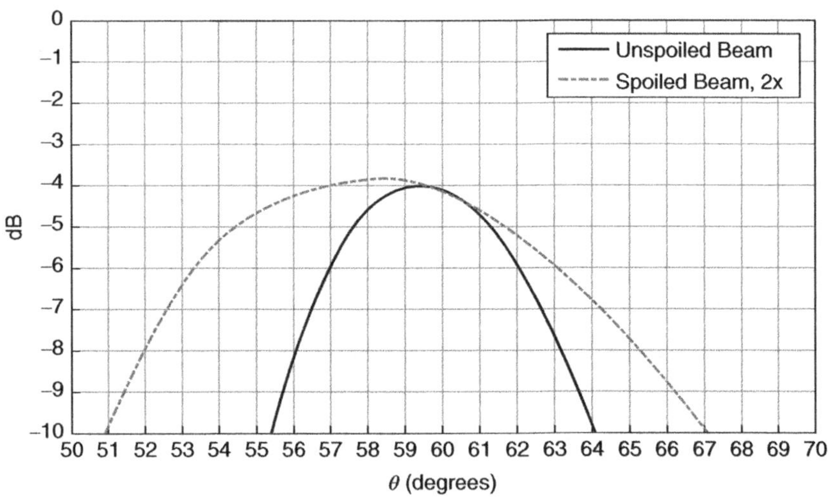

Figure 5.18 Zoomed-in view of the patterns shown in Figure 5.17. Both patterns are normalized to the own peak values at boresite ($\theta = 0$).

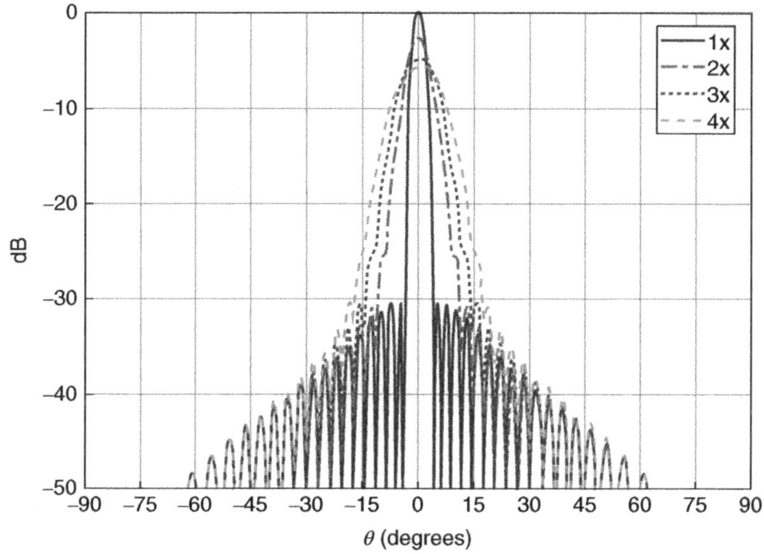

Figure 5.19 2x, 3x, and 4x beam broadening shown using quadratic phase weighting.

their own peaks and did not show the loss that is incurred with beam spoiling (nothing is for free!). In actuality, as the beamwidth is increased, the array gain is decreased, resulting in a loss in *ERP* or $\frac{G}{T}$. This loss has to be taken into consideration at the system level and budgeted. Figure 5.20 shows a zoomed-in view of the main beam region of Figure 5.19. Additionally, the quadratic phases that are used to generate the patterns in Figure 5.19 are also shown in Figure 5.20. The beam spoiling comes at a price with 3 dB, $\sim$ 5 dB, and $\sim$ 6 dB of loss for 2x, 3x, and 4x beam broadening, respectively. The loss shown in Figure 5.20 intuitively makes sense. An array without beam spoiling that has half the size would have 3 dB less gain. Similarly, this applies for a 3x and 4x reduction in the array size without beam spoiling, which would result in $\sim$ 5 dB and $\sim$ 6 dB of loss, respectively. Physics wins!

5.6 Monopulse for Angle Estimation

Monopulse systems provide angle error and measure angle of arrival (AoA) relative to the array on a single pulse [Hovanessian, 1984]. AESAs are used to employ monopulse for radars to search and detect multiple objects [Skolnik, 1990] and are also used to provide AoA for geolocation of emitter signals. Monopulse systems are also essential for high-precision tracking radars.

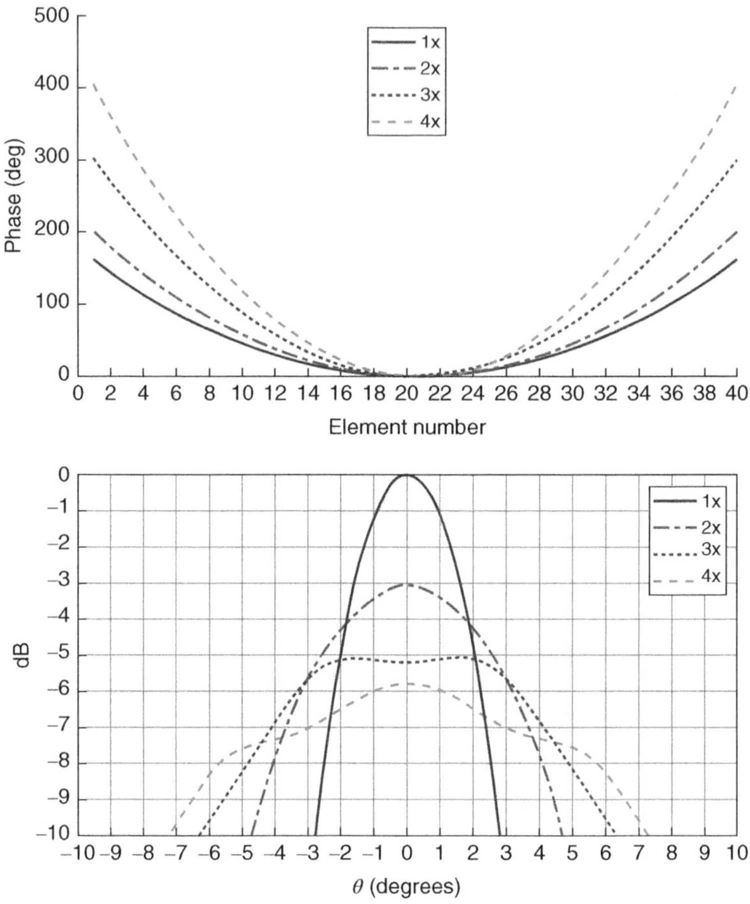

Figure 5.20 Quadratic phases shown in addition to a zoomed-in view of the patterns in Figure 5.19.

5.6.1 Three Channel Monopulse with an AESA

The AESA can be divided into quadrants to form a sum beam that is the coherent sum of the all the elements and two *delta* beams, which both have a null at the commanded scan angle in a single dimension. Figure 5.21 shows a pattern cut of a sum beam and a delta beam. The delta azimuth beam is formed by applying a 180° phase shift between the left and right halves of the AESA. The delta elevation beam is formed similarly except the 180° phase shift is applied between the top and bottom halves of the array.

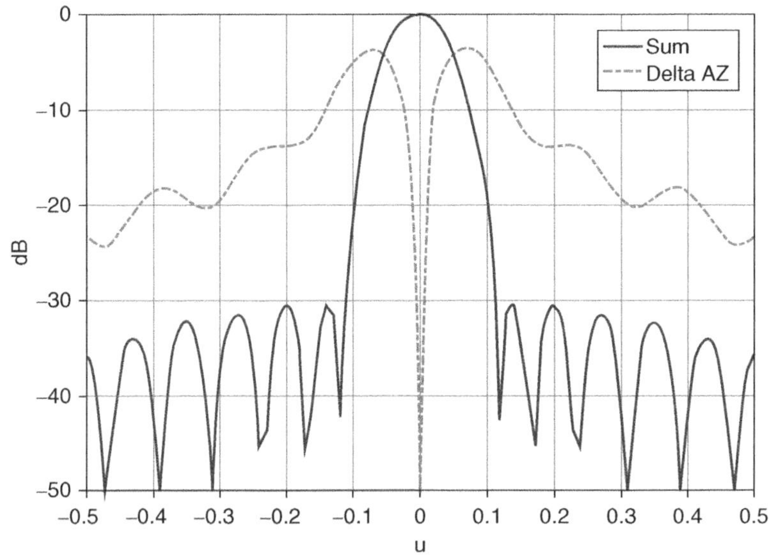

Figure 5.21 Pattern cut examples for a sum and delta azimuth beam. The delta azimuth beam has a null where the sum beam has a peak value.

Three separate beamformers could be used to form the monopulse sum and delta beams; however, this is not required. An alternative approach is to use the same beamformer to generate the sum and delta beams. The monopulse beams can be formed using a single beamformer with 180° hybrid couplers that use the quadrants of the AESA to form the monopulse beams. This is illustrated in Figure 5.22. 180° hybrid couplers provide a 180° phase shift between its two output ports [Balanis, 1982]. This enables a sum, delta azimuth, and delta elevation beam to be formed and then sent to the receiver for processing. The fourth output shown in Figure 5.22 is usually terminated with a resistive load, but it can also be used as an auxiliary beam to compare with the main beam (sum) for rejecting interference due to jammers or unintentional interference [Skolnik, 1990]. If the delta delta beam is used, an additional receiver channel is required. Figures 5.23 and 5.24 show the 2D patterns for the sum and delta beams. The relationship between an auxiliary beam and a sum beam is elaborated on further in Appendix F.

The ratio of the delta-to-sum beam is used to calculate the AoA of the incident energy. This ratio is referred to as an S curve since the ratio has the appearance of an S or backward S (see Figure 5.25). Assuming the AESA in Figure 5.22 has

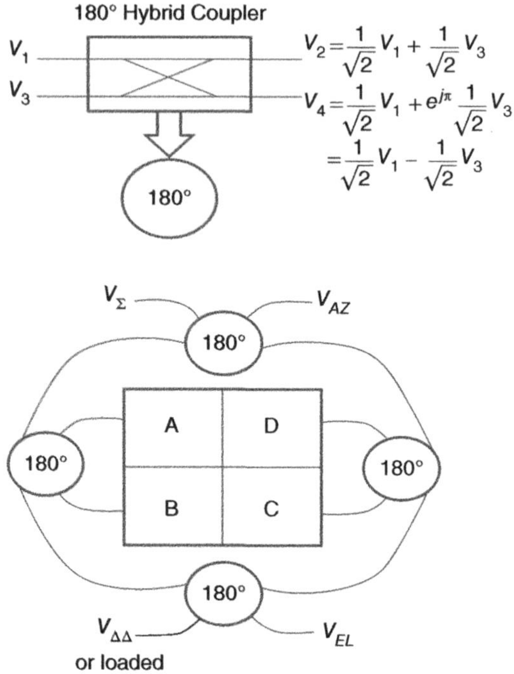

Figure 5.22 Using 180° hybrid couplers enables the sum and delta beams to be generated with a single beamformer. The fourth port can be either loaded or used as a delta delta beam (Figure 5.23) for sidelobe blanking.

length of L_x in the x dimension and a length of L_y in the y dimension, the voltage at the outputs of the quadrants can be represented as

$$V_i = AF \cdot e^{jX_i \frac{2\pi}{\lambda}(u-u_o)} \cdot e^{jY_i \frac{2\pi}{\lambda}(v-v_o)}, \; i = A,\; B,\; C,\; or\; D. \tag{5.24}$$

The X_i and Y_i are simple the phase center of the quadrants in Figure 5.22 and assuming the center of the AESA is (0,0) are

$$\begin{aligned}(X_A, Y_A) &= \left(\frac{-L_x}{4}, \frac{+L_y}{4}\right),\\ (X_B, Y_B) &= \left(\frac{-L_x}{4}, \frac{-L_y}{4}\right),\\ (X_C, Y_C) &= \left(\frac{+L_x}{4}, \frac{-L_y}{4}\right),\\ (X_D, Y_D) &= \left(\frac{+L_x}{4}, \frac{+L_y}{4}\right).\end{aligned} \tag{5.25}$$

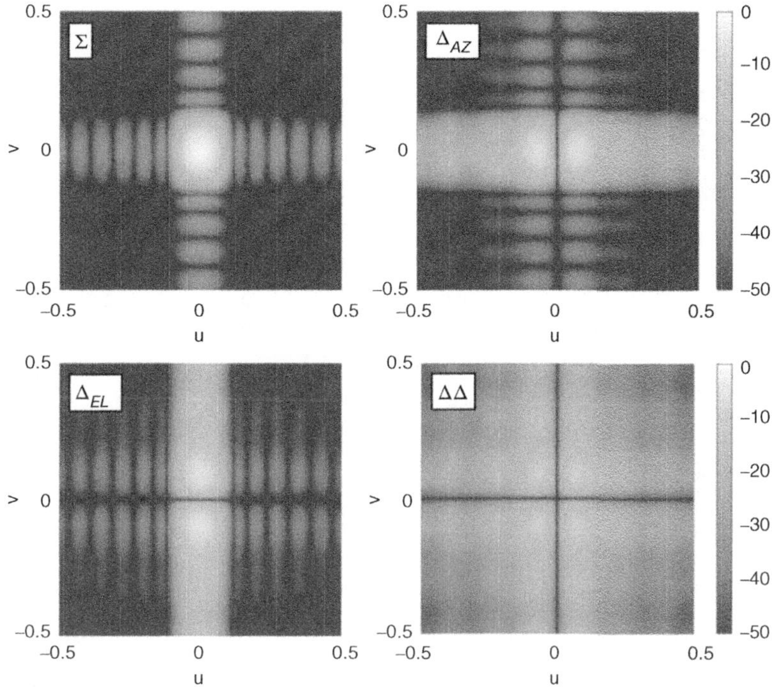

Figure 5.23 Sum, delta azimuth, delta elevation, and delta delta patterns generated by the monopulse beamformer shown in Figure 5.22.

The sum (Σ), delta azimuth (Δ_{AZ}), and delta elevation (Δ_{EL}) outputs can then be expressed as

$$
\begin{aligned}
\Sigma &= V_A + V_A + V_A + V_A \\
&= 4AF\ cos(\frac{L_x}{4}k\Delta u)cos(\frac{L_y}{4}k\Delta v) \\
\Delta_{AZ} &= (V_A + V_B) - (V_C + V_D) \\
&= -4jAF\ sin(\frac{L_x}{4}k\Delta u)cos(\frac{L_y}{4}k\Delta v) \\
\Delta_{EL} &= (V_A + V_D) - (V_B + V_C) \\
&= +4jAF\ cos(\frac{L_x}{4}k\Delta u)sin(\frac{L_y}{4}k\Delta v)
\end{aligned}
\tag{5.26}
$$

where $\Delta u = u - u_o$ and $\Delta v = v - v_o$. Also, the AF for each quadrant is assumed to be the same. For general performance this is sufficient, but to accurately model the performance with element amplitude and phase errors, a different AF needs to be used for each quadrant. Using Equation 5.26, the S ratio expressions can be

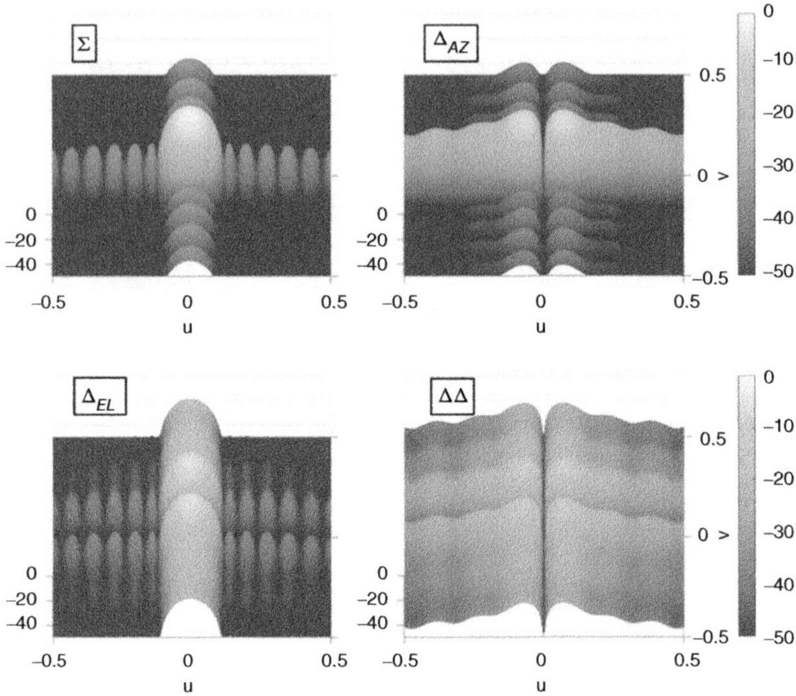

Figure 5.24 Isometric views of the three channel monopulse beam patterns shown in Figure 5.23.

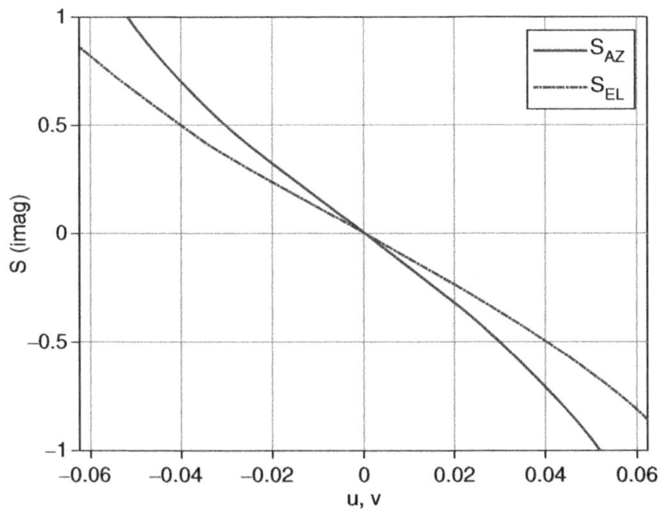

Figure 5.25 The S curve is used to determine the AoA for received signals incident on the AESA.

written as

$$S_{\Delta_{AZ}} = \frac{\Delta_{AZ}}{\Sigma} = -jtan\left(\frac{L_x}{4}k\Delta u\right),$$
$$S_{\Delta_{EL}} = \frac{\Delta_{EL}}{\Sigma} = +jtan\left(\frac{L_y}{4}k\Delta v\right). \quad (5.27)$$

By taking the imaginary value of the measured S ratio, the AoA can be calculated.

5.6.1.1 Calibration for Monopulse Coupler Errors

The 180° hybrid couplers shown in Figure 5.22 are a source of amplitude and phase error that can degrade AESA monopulse AoA performance. Hybrid coupler errors cause a translation and rotation of the monopulse S-curve. These errors can be calibrated by measuring/characterizing the amplitude and phase error offsets and storing them in a table. These values can then be applied to the measured S ratio for correction.

The coupler used to form the Σ and Δ_{AZ} beams will be used to illustrate the effects of coupler errors. A similar formulation can be done for the Δ_{EL} beam. Using Equation 5.27, the measured S ratio after the final coupler stage can be represented as

$$\tilde{S}_{\Delta_{AZ}} = \frac{\tilde{\Delta}_{AZ}}{\tilde{\Sigma}} = \frac{\gamma\Delta_{AZ}}{\epsilon\Sigma}, \quad (5.28)$$

where γ and ϵ are the complex errors associated with the Σ and Δ_{AZ} beams, respectively. The expression in Equation 5.28 can be rearranged as shown below:

$$\begin{aligned}\tilde{S}_{\Delta_{AZ}} &= \frac{\gamma\Delta_{AZ}}{\epsilon\Sigma} \\ &= \frac{|\gamma|e^{j\phi_\gamma}}{|\epsilon|e^{j\phi_\epsilon}} \cdot \frac{\Delta_{AZ}}{\Sigma} \\ &= \frac{|\gamma|}{|\epsilon|} \cdot e^{j(\phi_\gamma-\phi_\epsilon)} \cdot \frac{\Delta_{AZ}}{\Sigma} \\ &= \delta_{amp} \cdot e^{j\delta_{phase}} \cdot \frac{\Delta_{AZ}}{\Sigma}. \end{aligned} \quad (5.29)$$

The amplitude and phase error shown in Equation 5.29 can affect the AoA error and must be calibrated. Typically in an AESA, the Σ beam is tuned to maximize the AESA antenna gain. This means that the Δ beam that shares the same coupler as the Σ beam will be optimized and calibrated also. The other Δ beam will not be optimized and may suffer some degradation in accuracy.

5.6.2 Two-Channel Monopulse with an AESA

The monopulse approach previously discussed assumes that there are three receiver channels dedicated for Σ, Δ_{AZ}, and Δ_{EL} AESA beam outputs. In most modern AESAs this is a valid assumption. However, it is possible to perform monopulse with the AESA using only two channels. The three-channel approach uses an amplitude comparison in order to determine AoA. A two-channel approach, which will be referred to here as radial monopulse, requires both amplitude and phase comparison to determine AoA. Figure 5.26 illustrates the

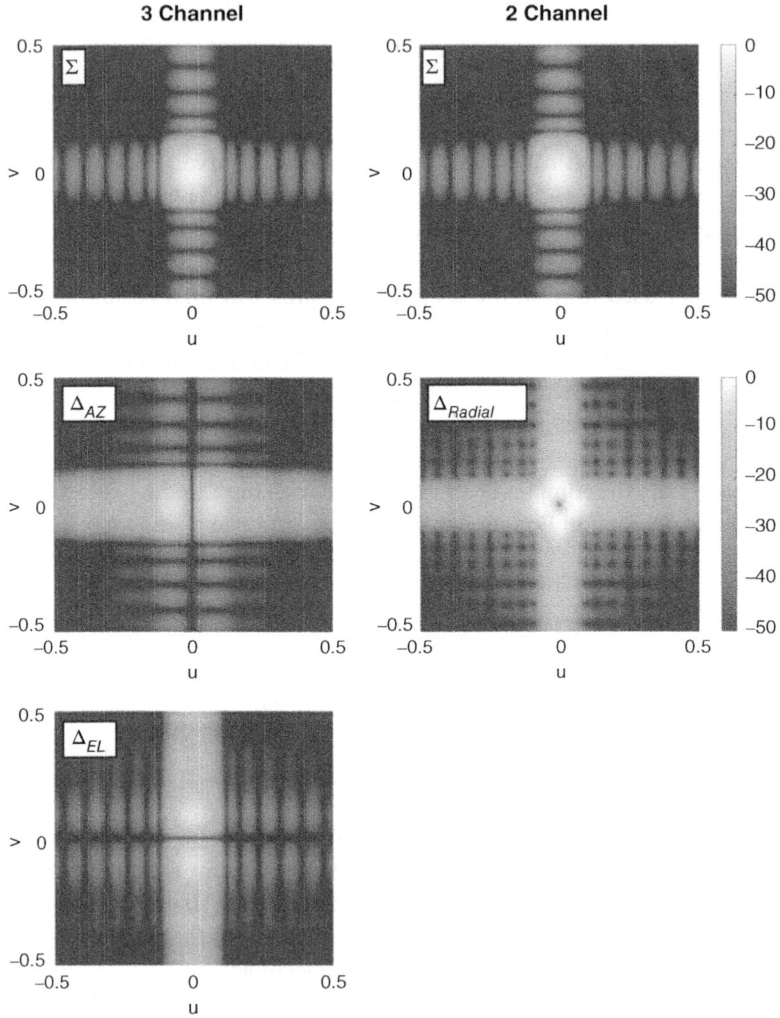

Figure 5.26 The three-channel approach uses an amplitude comparison for AoA, while the two-channel, or radial monopulse, approach requires both amplitude and phase comparison for AoA. Similar to the Δ_{AZ} and Δ_{EL} beams, the Δ_{Radial} beam has a null at the peak of the Σ beam. The difference is that it is a radial null.

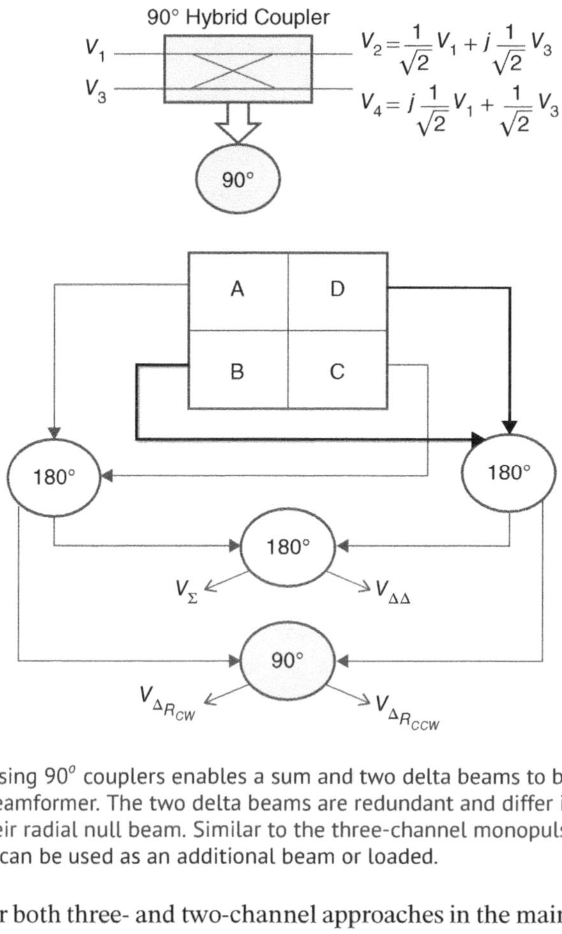

Figure 5.27 Using 90° couplers enables a sum and two delta beams to be generated with a single beamformer. The two delta beams are redundant and differ in the clock angle of the their radial null beam. Similar to the three-channel monopulse beamformer, the fourth port can be used as an additional beam or loaded.

2D patterns for both three- and two-channel approaches in the main beam region. Radial monopulse uses a delta beam that has a donut-shaped annular ring with a null at its center. The Δ-to-Σ ratio, similar to a three-channel monopulse, is then used to determine AoA.

Figure 5.27 illustrates how the Σ and Δ beams are formed. A 90° hybrid coupler is used in addition to 180° hybrid couplers. The output of the 90° hybrid coupler generates two Δ beams: clockwise (CW) and counterclockwise (CCW). This means that the phase of the delta beams increases monotonically in either the CW or CCW direction. Figure 5.28 shows the magnitude and phase of both the Σ and Δ_{CCW} beams. The phase of the CCW beam has a clocked phase. The CW beam (not shown) has the same behavior, but the phase increases in the CW direction. The S ratio can be expressed as

$$S_{R_{CCW}} = \frac{\Delta_{R_{CCW}}}{\Sigma} = |S_{R_{CCW}}| e^{j\Phi_{CCW}}. \tag{5.30}$$

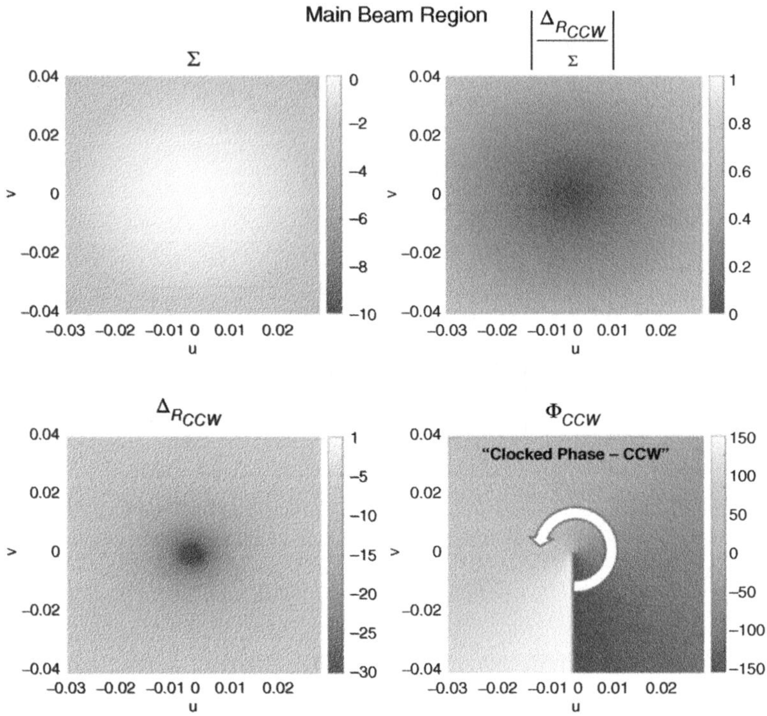

Figure 5.28 Sum beam and CCW radial null beam shown in addition to the amplitude and phase ratio between the two. The amplitude and phase uniquely define the AoA.

In Equation 5.30, the amplitude places the AoA on a contour around the main beam, and the phase determines where on the contour the AoA resides. The two-channel approach reduces the number of receiver channels required but is not as accurate as the three-channel method described earlier.

5.6.2.1 Low Sidelobe Delta Beams

The S ratio shown in Equation 5.27 creates an S-Curve as shown in Figure 5.25. The steeper the S-curve, the more accurate the AoA will be when measured. Using a standard delta beam as previously shown does not provide the optimum AoA. To improve this, a low-sidelobe delta beam was developed [Bayliss, 1968]. The Bayliss distribution has low SLs, which help to create a sharper delta beam, which in turn sharpens the S-curve. The beamformer with Bayliss weighting is typically a separate beamformer from the sum beamformer and is not constructed as described for the three-channel approach covered earlier.

References

Balanis, C. *Antenna Theory Analysis and Design*. John Wiley & Sons, Publishers, Inc., 1982.

Bayliss, E.T. "Design of monopulse antenna difference patterns with low sidelobes." *Bell System Technical Journal*, pp. 623–640, 1968.

Brown, G., Kerce, J., and Mitchell, M. "Extreme beam broadening using phase only pattern synthesis." *4th IEEE Workshop on Sensor Array and Multichannel*, pp. 36–39, 2006.

Holzman, E.L., and Agrawal, A.K. "A comparison of active phased array, corporate beamforming architectures." *Proceedings of International Symposium on Phased Array Systems and Technology*, pp. 429–434, 1996.

Hovanessian, S.A. *Radar System Design and Analysis*. Artech House, 1984.

Hu, C., Chuang, C., and Hung, C. "Phase-only pattern synthesis for the design of an active linear phased array." *Proceedings 2000 IEEE International Conference on Phased Array Systems and Technology*, pp. 275–278, 2006.

Lin, S.-M., Wang, Y.-Q., and Shen, P.-L. "Phase-only synthesis of the shaped beam patterns for the satellite planar array antenna." *Proceedings of IEEE International Conference on Phased Array Systems and Technology*, pp. 331–334, 2000.

Mailloux, R. "Antenna array architecture." *Proceedings of the IEEE*, pp. 163–172, 1992.

Mailloux, R.J. *Phased Array Antenna Handbook*. Artech House, Norwood, MA, 1993.

Sayidmarie, K.H., and Sultan, Q.H. "Synthesis of wide beam array patterns using quadratic-phase excitations." *International Journal of Electromagnetics and Applications*, pp. 127–135, 2013.

Skolnik, M.I. *Radar Handbook*. McGraw Hill, 1990.

Taylor, T.T. Design of line-source antennas for narrow beamwidth and low side lobes. *Transactions of the IRE Professional Group on Antennas and Propagation*, 3:16–28, 1955. doi: 10.1109/TPGAP.1995.5720407.

6

AESA Cascaded Performance

Key Concepts

- AESA Output Signal and Noise Power
- AESA Signal/Noise Gain and Noise Factor
- AESA *nth*-Order Intercept Point (IP_n)
- AESA Spurious Free Dynamic Range (SFDR)

6.1 Introduction

An AESA is composed of three major components: the array elements, TRMs, and beamformer. All three add their own complexity to the AESA design, in addition to the integration involved to make the AESA operational. Even with this complexity, AESAs can be represented as active devices with equivalent parameters that represent the performance of the AESA as a whole. In Figure 6.1 the AESA is represented as a single device with equivalent directive gain , G_{AESA}, electronic or active gain $G_{a_{AESA}}$, and noise factor F_{AESA} with input and output signal and noise power (S_{in}, N_{in}, S_{out}, and N_{out}). The directive gain only affects the signal power and represents the antenna directivity. The electronics gain affects both signal and noise power and is a function of the active device gain (amplifiers) and attenuation in the AESA. The equivalent noise factor represents the added noise of the AESA, which affects the output SNR and dynamic range.

Figure 6.1 can be used to represent both transmit and receive operations. This chapter will focus on receive, but the same principles can be used for transmit

Active Electronically Scanned Arrays: Fundamentals and Applications, First Edition. Arik D. Brown.

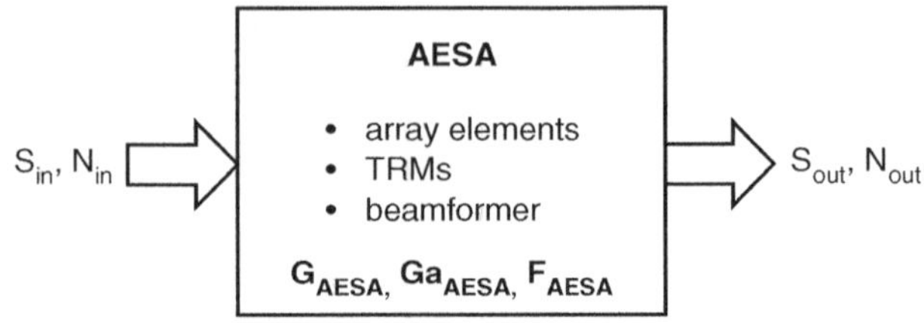

Figure 6.1 An AESA can be represented with equivalent gain and noise factor parameters that summarize its performance similar to an active device.

operation. Typically, on transmit, as discussed in Chapter 5, there is uniform weighting and the cascaded transmit power is straightforward to calculate. Additionally, for transmit the primary system parameter is ERP, which is also captured by the cascaded transmit power calculation. For receive operation, as discussed in Chapters 1 and 4, $\frac{G}{T}$ and linearity are the primary drivers for performance and must be calculated correctly to properly characterize the system performance. It is essential to understand how to properly calculate the receive cascaded performance, which in most cases is much more complicated than the transmit case. In receive operation there is nominally some level of amplitude weighting used, which can be distributed between both the TRMs and beamformer and so affects the output *SNR*.

The block diagram shown in Figure 6.2 will be used throughout this chapter to derive key AESA system performance parameters. In Figure 6.2 signal and noise enter each of the array elements. S_{in_m} and N_{in_m} are placed on the right of the array element to encompass the loss due to the element for both S_{in_m} and N_{in_m}. This also includes the loss due to the radome. These losses are prior to the first amplifier in the receive chain (*LNA*) and must be accounted for because they are a one-for-one effect on increasing the noise figure, which ultimately reduces the output *SNR*.

Equation 6.1 details the expressions for both S_{in_m} and N_{in_m} shown in Figure 6.2.

$$\begin{aligned} S_{in_m} &= \frac{ERP}{4\pi R^2} \cdot A_e \cdot L_{array\ element} \\ &= \frac{ERP}{4\pi R^2} \cdot \frac{\lambda^2 D_e \cos^{EF}(\theta)}{4\pi} \cdot L_{array\ element}, \\ N_{in_m} &= kTB \cdot L_{array\ element}. \end{aligned} \tag{6.1}$$

The ERP shown in Equation 6.1 represents either the ERP from an external emitter ($P_{TX_{ext}} G_{ext}$) or reflected target signal power ($\frac{P_{TX} G_{TX}}{4\pi R^2} \cdot \sigma$). The array element loss, $L_{array\ element}$ is assumed to be the same for every element, and D_e is the element directivity of a single array element. Finally, the noise temperature T represents the noise temperature from the environment as seen by the AESA. In someapplications T can be set to T_o, but here it will be left as T as a more general representation. For the formulations to follow, S_{in_m} and N_{in_m} will be used for simplification.

After the array elements in Figure 6.2, the signal and noise travel through the TRMs that have an electronic gain G_{a_m}, noise factor F_m, and attenuation loss β_m. As shown in Chapter 4, the TRM receive chain is composed of various components consisting of a limiter, LNA, phase shifter, switches, line length losses, and in some cases additional amplifiers. For simplicity, it will be assumed that the TRM has an amplifier and an attenuator. This is valid because all of the components described in the receive chain can be combined into an effective gain. This representation is highlighted in Figure 6.3, where the TRM is composed of an amplifier and an attenuator. It is important to note that β_m is an attenuation amplitude weight applied to the signal voltage. The power loss associated with this attenuator is β_m^2.

The final stage in the Figure 6.2 block diagram is the beamformer. Exactly as shown in Chapter 5, the weights in the beamformer are designated as α_m. The output signal and noise power from the beamformer are sent to the downconverters and receiver for the system.

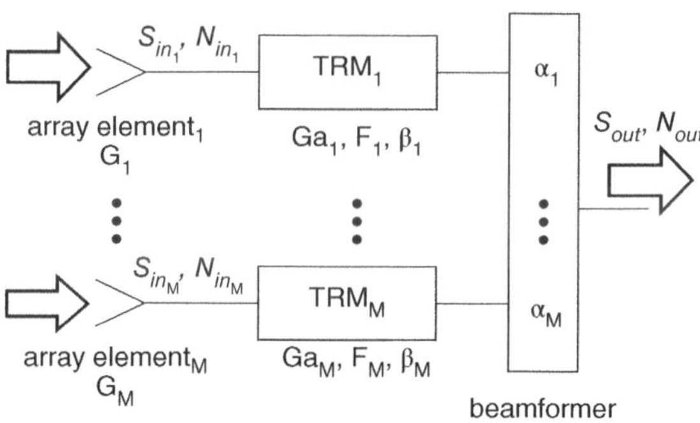

Figure 6.2 Reference block diagram for the AESA that will be used to derive system level performance parameters.

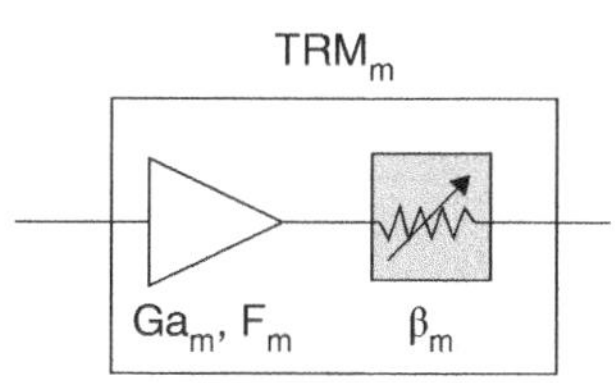

Figure 6.3 The TRM can be represented with an equivalent electronic gain, G_{a_m}, noise factor, F_m, and attenuation, β_m.

6.2 Fundamental Expressions for Cascade Calculations

Before deriving the formulation for the output signal and noise power, a few fundamental expressions will be reviewed. These expressions will be used to represent the cascaded noise power in the AESA.

6.2.1 Noise Model

Calculating the noise gain of a device provides the foundation for determining the noise gain of a series of cascaded devices. This type of analysis is performed for any RF circuit chain and provides the basis for quantifying system performance in terms of *SNR*. The cascaded equation for noise gain will enable the calculation of the *SNR* for an AESA.

6.2.1.1 Active Device

Figure 6.4 illustrates an equivalent model for the added noise of an active device. The top graphic shows that the resultant output noise power, N_{out}, is the input noise multiplied by the electronic gain, G_a, and the noise factor, F_a, of the device. The noise factor is a measure of how much self-generated noise the device adds to the input noise [Pettai, 1984]. The noise factor has a minimum value of 1, which is equivalent to $N_{out} = N_{in}$ and no self-generated noise ($F_a = 1$). In reality, every active device has its own self-generated noise, and F_a is always greater than 1.

The equivalent model in Figure 6.4 shows that the device's self-generated noise is *additive*. The cumulative noise at the device's output is the sum of the amplified input noise and the amplified self-generated noise. This relation will be useful when calculating the equivalent noise factor for devices cascaded in series.

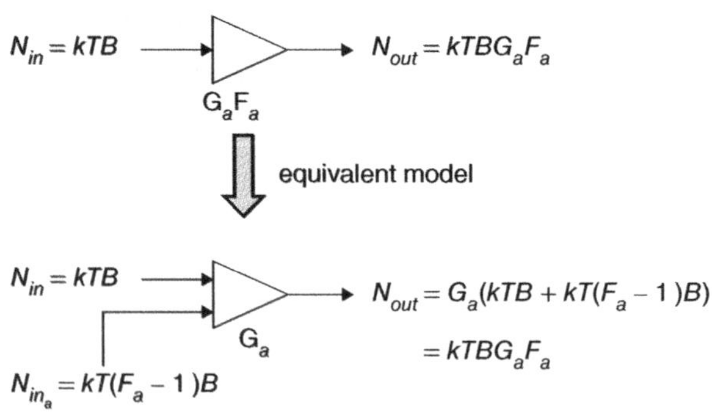

Figure 6.4 Equivalent noise model for an amplifier. The equivalent formulation shows that the noise is additive.

6.2.1.2 Resistive Device

The noise factor for a resistive device such as an attenuator is calculated differently than that of an active device. This is shown in Figure 6.5. As previously mentioned, the loss of an attenuator is equal to β^2, where β^2 is the resistive power loss. β^2 has a maximum value of 1.

6.2.1.3 Noise Factor Definition

The definition of noise factor is the ratio of the input *SNR* to the output *SNR* [Pettai, 1984]. This expression is shown in Equation 6.2.

$$F = \frac{\frac{S_{in}}{N_{in}}}{\frac{S_{out}}{N_{out}}} = \frac{S_{in}}{S_{out}} \cdot \frac{N_{out}}{N_{in}}. \tag{6.2}$$

There are two important points about F that have their origin in Equation 6.2. First, by increasing S_{out}, F is minimized. Since S_{out} is a function of the AESA antenna directivity, F is minimized by maximizing the directivity. Second, minimizing N_{out} minimizes F. This means that it is desirable to minimize the self-generated noise in order to decrease F.

Having established the definition for F, it is useful to revisit the noise models for the active and resistive devices, and calculate F based on the definition in Equation 6.2. Equation 6.3 shows the calculation of the noise factor for an active device:

$$F = \frac{\frac{S_{in}}{N_{in}}}{\frac{S_{out}}{N_{out}}} = \frac{\frac{S_{in}}{kTB}}{\frac{G_a S_{in}}{kTBG_a F_a}} = F_a. \tag{6.3}$$

From Equation 6.3 it is shown that the noise factor for the active device is F_a. Similarly, the noise factor for the resistive device can be calculated using Equation 6.4.

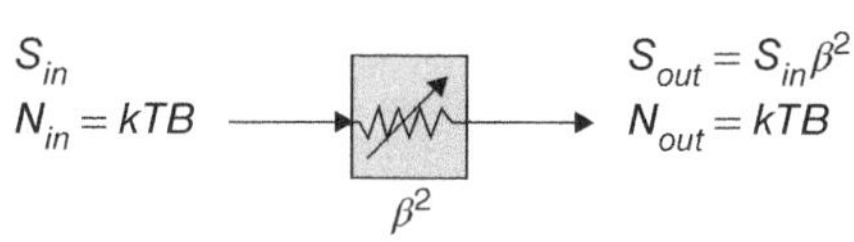

Figure 6.5 Equivalent noise model for a resistor.

$$F = \frac{\frac{S_{in}}{N_{in}}}{\frac{S_{out}}{N_{out}}} = \frac{\frac{S_{in}}{kTB}}{\frac{\beta_a^2 S_{in}}{kTB}} = \frac{1}{\beta_a^2}. \tag{6.4}$$

For a resitive device the output noise floor is not increased, but the signal power is decreased by β_a^2, and therefore the *SNR* is decreased by β_a^2. This means that the noise figure, $10log_{10}\left(\frac{1}{\beta_a^2}\right)$, of the resistor is simply the negative of the resistor gain in dB, $-10log_{10}\left(\beta_a^2\right) = 10log_{10}\left(\frac{1}{\beta_a^2}\right)$.

6.2.2 Cascaded Noise Factor

The noise models for active and resistive devices can be used to model the cascade of devices in series. The electronic signal gain for cascaded devices is straightforward and is simply the multiplicative gain (positive or negative) of each device in the cascade. The net electronic signal gain is then the product of all of the device gains. The cascaded noise gain can be calculated the same way; however, it will be derived by calculating the cascaded noise factor. Using the noise models it will be shown that the added self-generated noise from each device contributes to output noise gain of the final device. An equivalent noise gain formula for the entire cascade can then be derived.

To illustrate the calculation of the cascaded electronic signal and noise gains, a two-device example will be used followed by the complete formula expression for N devices in series. Figure 6.6 shows two amplifiers in series, each with a different electronic gain (G_{a_1}, G_{a_2}) and noise factor (F_1, F_2). For clarity, it is again noted that the subscript a for the electronic gain denotes the gain of the device and is not antenna gain/directivity. As previously mentioned, for the two cascaded

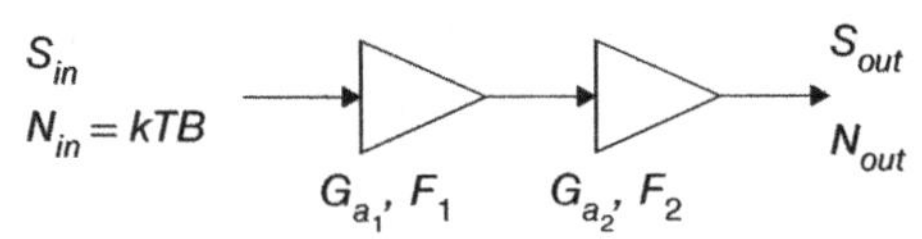

Figure 6.6 Two amplifiers cascaded used as an example to calculate the cascaded signal and noise gains.

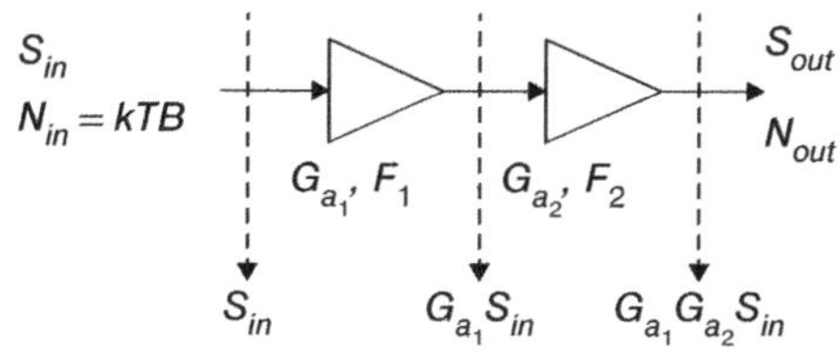

Figure 6.7 Cascaded signal gain $G_{a_{21}}$ for the two device example is $G_{a_1}G_{a_2}$.

devices, S_{out} is the product of the electronic signal gains and S_{in}. This is illustrated in Figure 6.7 and expressed in Equation 6.5.

$$\begin{aligned} S_{out} &= G_{a_1}G_{a_2}S_{in} \\ &= G_{a_{21}}S_{in}, \ \ G_{a_{21}} = G_{a_1}G_{a_2} \end{aligned} \tag{6.5}$$

To calculate the noise power gain for the two devices shown in Figure 6.6, the equivalent noise model from Figure 6.4 can be used. Substituting this equivalent model into Figure 6.6 results in an equivalent representation of the two cascaded devices. This is shown in Figure 6.8. This representation provides a straightforward method for calculating N_{out}. At the output of the first amplifier the noise power is $kTF_1BG_{a_1}$, as shown in Figure 6.9. The output of the second amplifier is the sum of the output noise of the first amplifier and the self-generated noise of the second amplifier scaled by the amplifier gain.

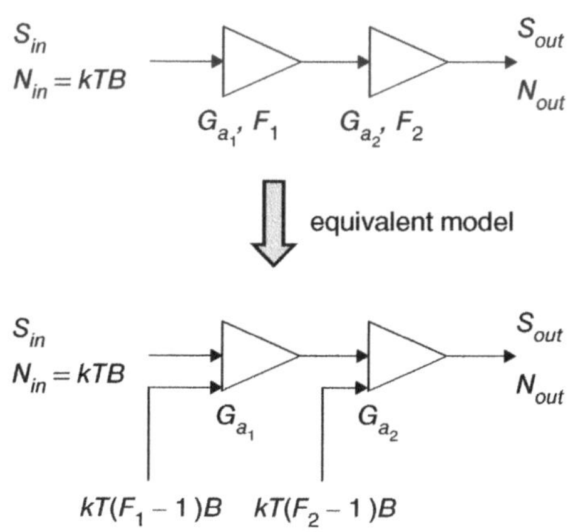

Figure 6.8 Equivalent cascaded model used to calculate the cascaded electronic gain and noise factor. These parameters are required to compute the output noise power.

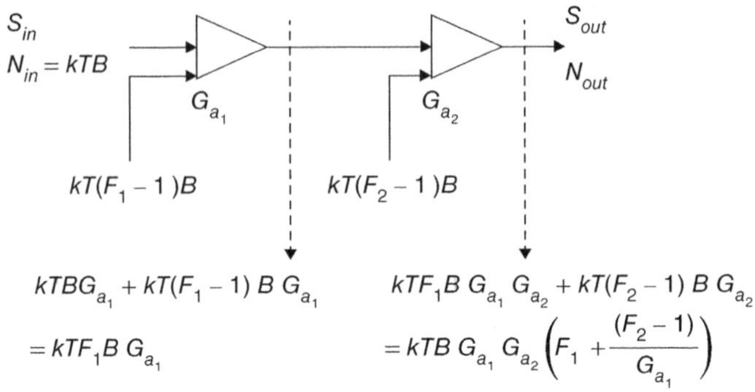

Figure 6.9 Using the cascaded noise model, the cascaded output noise power is $kTB\left(G_{a_1}G_{a_2}\right)\left(F_1+\frac{(F_2-1)}{G_{a_1}}\right)$.

N_{out} can then be expressed as

$$\begin{aligned} N_{out} &= kTB\left(G_{a_1}G_{a_2}\right)\left(F_1+\frac{(F_2-1)}{G_{a_1}}\right) \\ &= kTBG_{a_{21}}F_{21}. \end{aligned} \tag{6.6}$$

$G_{a_{21}}$ and F_{21} are the cascaded electronic gain and noise factor for the two amplifiers. This is further illustrated in Figure 6.10. From Equation 6.6, it is seen that if

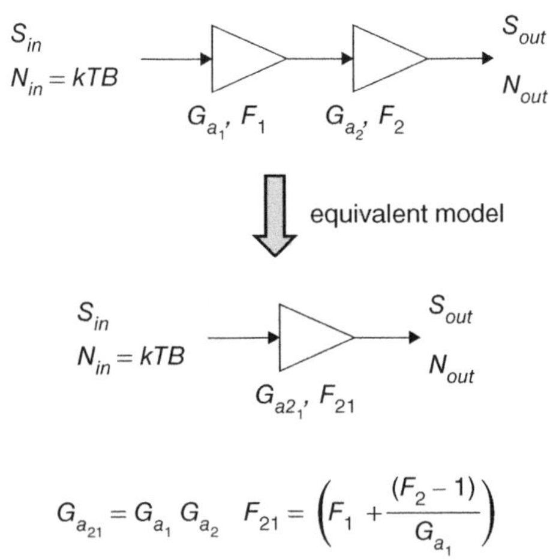

Figure 6.10 The cascaded devices can be represented by a single device with equivalent electronic gain and noise factor.

the electronic gain of the first amplifier is made large where the second term for the cascaded noise factor, $\frac{(F_2-1)}{G_{a_1}}$, is much smaller than F_1, then the self-generated noise of the second amplifier will be minimized. This holds true for a cascade of N devices also and is why balancing the noise power gain is essential for optimizing *SNR*.

Figure 6.11 shows a cascade of N amplifiers in series. The resulting noise power gain for N devices is then

$$N_{out} = kTBG_{a_1} \cdots G_{a_N} \left(F_1 + \frac{(F_2 - 1)}{G_{a_1}} + \cdots + \frac{(F_N - 1)}{G_{a_1} \cdots G_{a_{N-1}}} \right) \tag{6.7}$$

Although amplifiers are shown in Figure 6.11, Equation 6.7 still holds for stages in the cascade that are resistive devices. This will be made apparent in the calculation of the AESA output noise power. From Equation 6.7 the cascaded electronic gain, noise factor, and noise temperature can be expressed as

$$\begin{aligned} G_{a_{cascasde}} &= G_{a_1} \cdots G_{a_N}, \\ F_{cascade} &= \left(F_1 + \frac{(F_2 - 1)}{G_{a_1}} + \cdots + \frac{(F_N - 1)}{G_{a_1} \cdots G_{a_{N-1}}} \right), \\ T_{cascade} &= TF_{cascade} \\ &= T\left(F_1 + \frac{(F_2 - 1)}{G_{a_1}} + \cdots + \frac{(F_N - 1)}{G_{a_1} \cdots G_{a_{N-1}}} \right). \end{aligned} \tag{6.8}$$

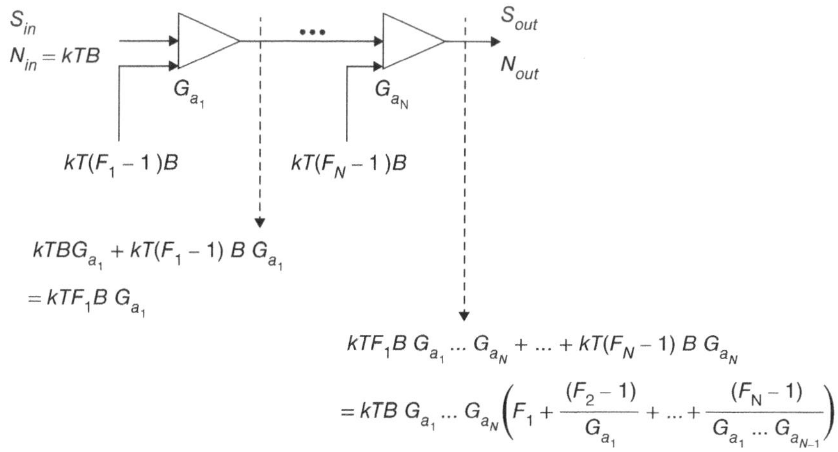

Figure 6.11 Building upon Figure 6.9 the same methodology is used to calculate the equivalent noise factor for N cascaded devices.

6.3 AESA Cascaded Performance

Having defined the formulas for cascaded gain and noise factor, these formulations can now be applied in order to calculate them for an AESA. The starting point will be Figure 6.2. In order to calculate the cascaded parameters, typically a spreadsheet is created that tracks the gain and noise factor for every component in a single array element channel. Using the formulation for a lossless beamformer derived in Chapter 5, the coherent gain of any combiner or beamformer section is accounted for by applying a gain for the signal gain and not the noise gain. Ohmic losses are applied to the signal and noise gain. The single string representation is shown in Figure 6.12. Although the TRM is simplified, this representation remains valid. The cascaded equations take into account the gains and noise factors of any additional components internal or external to the TRM and can be modified to include these components for an actual AESA design.

6.3.1 AESA Output Signal Power

As shown earlier for the two-amplifier example, the signal gain is simply the cascade of the gains in the AESA chain. It is important to point out that the term *signal gain* is being used and not *electronic gain*. This is because the signal gain is composed of both directive gain from the array elements (Equation 6.1) and electronic gain from the active devices.

For the AESA signal gain and later the AESA noise gain, the gain will be traced through the block diagram in Figure 6.12. Starting with the signal gain, the signal power after the amplifier stage in TRM_m is

$$S_{m_{postamplifier}} = G_{a_m} S_{in_m}. \tag{6.9}$$

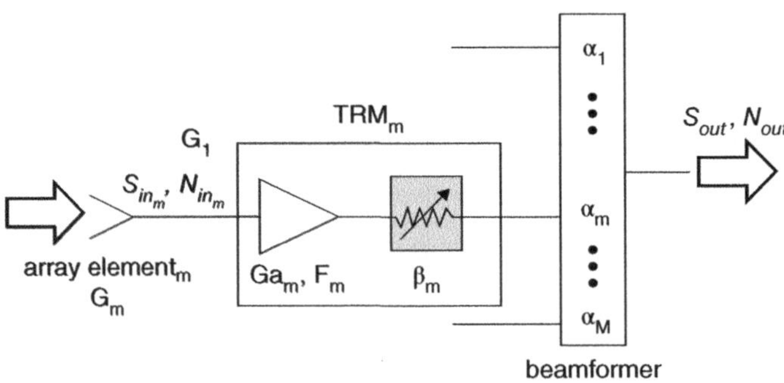

Figure 6.12 Single string model used to calculate the cascaded signal and noise parameters for an AESA.

After the TRM resistor, the signal power is

$$S_{m_{postattenuator}} = \beta_m^2 G_{a_m} S_{in_m}. \tag{6.10}$$

After the beamformer, the resultant output signal power is

$$S_{beamformer} = \left| \sum_{m=1}^{M} \alpha_m \sqrt{\beta_m^2 G_{a_m} S_{in_m}} \right|^2. \tag{6.11}$$

The square root in Equation 6.11 is required because the beamformer transfer function, as shown in Chapter 5, is for voltage and not power. A logical question then is, since the voltages from the TRMs will have different phases, how is that phase taken into account? The assumption here is that the coherent addition of the voltages in the beamformer will provide a maximum or a reduced value depending on the location of the incoming signal in the main beam. The reduction is captured with the array element directive gain that is included in the term for S_{in_m} (see Equation 6.1). Additionally, the phase over the main beam of the AESA is a constant value, so the formulation in Equation 6.11 has no loss in generality. With additional manipulation, the equation for S_{out} can be written as

$$\begin{aligned} S_{out} &= S_{beamformer} \\ &= \left| \sum_{m=1}^{M} \alpha_m \sqrt{\beta_m^2 G_{a_m} S_{in_m}} \right|^2 \\ &= G_{a_m} S_{in_m} \left| \sum_{m=1}^{M} \alpha_m \beta_m \right|^2. \end{aligned} \tag{6.12}$$

6.3.2 AESA Output Noise Power

Using the cascaded noise factor equation previously derived, the AESA output noise power can be calculated using the representative architecture in Figure 6.12. The composite electronic gain and noise factor for TRM_m are:

$$\begin{aligned} G_{TRM} &= G_{a_m} \beta_m^2, \\ F_{TRM} &= F_m + \frac{\left(\frac{1}{\beta_m^2} - 1\right)}{G_{a_m}}. \end{aligned} \tag{6.13}$$

Using Equation 6.13, the output noise power from the TRM is

$$\begin{aligned} N_{m_{TRM}} &= kTBG_{TRM}F_{TRM} \\ &= kTBG_{a_m}\beta_m^2 \left(F_m + \frac{\left(\frac{1}{\beta_m^2} - 1\right)}{G_{a_m}} \right). \end{aligned} \tag{6.14}$$

Thus far, T has been used to denote the external noise temperature to the AESA. In some applications the external noise is different due to environmental conditions. In Appendix G this is considered with an example of how to calculate the noise temperature presented to the AESA by the environment.

Using the expression for $N_{m_{TRM}}$ in Equation 6.14, the output noise power of the beamformer can be calculated. It is the noise power sum of the noise at each input to the beamformer. This can be expressed as

$$\begin{aligned} N_{beamformer} &= \sum_{m=1}^{M} \alpha_m^2 N_{m_{TRM}} \\ &= \sum_{m=1}^{M} \alpha_m^2 kTBG_{a_m}\beta_m^2 \left(F_m + \frac{\left(\frac{1}{\beta_m^2} - 1\right)}{G_{a_m}} \right). \end{aligned} \tag{6.15}$$

The output noise power can then be expressed as

$$\begin{aligned} N_{out} &= N_{beamformer} \\ &= \sum_{m=1}^{M} \alpha_m^2 kTBG_{a_m}\beta_m^2 \left(F_m + \frac{\left(\frac{1}{\beta_m^2} - 1\right)}{G_{a_m}} \right) \\ &= kTB\sum_{m=1}^{M} \alpha_m^2 \beta_m^2 G_{a_m} \left(F_m + \frac{\left(\frac{1}{\beta_m^2} - 1\right)}{G_{a_m}} \right). \end{aligned} \tag{6.16}$$

In Equation 6.16, G_{a_m} and F_{TRM} can be removed from the summation since the assumption for the single string model assumes that these values are equal for every element. In actuality this a good representation since the gain variance of the TRMs for the AESA are kept within a tight window to ensure there is no loss of coherent gain in the main beam. Additionally, the electronic gain of the TRM, G_{a_m}, is large enough such that the attenuator loss β_m^2 is not a driver to the noise factor of the TRM. To illustrate this second point further, consider a TRM amplitude weight, β, that is 0.6. This makes the numerator of the second term in F_{TRM} equal to $\left(\frac{1}{\beta_m^2} - 1\right) = \left(\frac{1}{0.6^2} - 1\right) = 1.78$. Typically the electronic gain in a well designed TRM is 20 dB or greater, so assuming it is 20 dB for this example, $G_{a_m} = 10^{(20/10)} = 100$. The ratio of $\left(\frac{1}{\beta_m^2} - 1\right)$ to G_{a_m} is then $\frac{1.78}{100} = 0.018$, which will have a small effect on F_{TRM}.

6.3.3 AESA Signal/Noise Gain and Noise Factor

Now that expressions for both AESA signal and noise output power have been formulated, the cascaded signal and noise gains and noise factor can be determined. From Equation 6.12, the AESA cascaded signal gain is

$$\begin{aligned} G_{signal_{AESA}} &= G_{a_m} G_d \left| \sum_{m=1}^{M} \alpha_m \beta_m \right|^2 \\ &= G_{a_m} D_e cos^{EF}(\theta) \cdot L_{array} \left| \sum_{m=1}^{M} \alpha_m \beta_m \right|^2 , \end{aligned} \tag{6.17}$$

where G_d is the directive gain of the element, $D_e = \frac{4\pi A_e}{\lambda^2}$, EF is the element factor, and L_{array} is the pre-LNA array element losses (radome, ohmic loss, mismatch loss).

The cascaded AESA noise gain can be expressed as

$$G_{noise_{AESA}} = G_{a_m} \sum_{m=1}^{M} \alpha_m^2 \beta_m^2, \tag{6.18}$$

using the output noise power expression in Equation 6.16. The ratio of $G_{signal_{AESA}}$ to $G_{noise_{AESA}}$ shows an important relationship. This ratio is is shown in Equation 6.19

$$\begin{aligned} \frac{G_{signal_{AESA}}}{G_{noise_{AESA}}} &= \frac{G_{a_m} D_e cos^{EF}(\theta) \cdot L_{array} | \sum_{m=1}^{M} \alpha_m \beta_m |^2}{G_{a_m} \sum_{m=1}^{M} \alpha_m^2 \beta_m^2} \\ &= \left(D_e cos^{EF}(\theta) \cdot L_{array} \right) \frac{\left| \sum_{m=1}^{M} \alpha_m \beta_m \right|^2}{\sum_{m=1}^{M} \alpha_m^2 \beta_m^2} . \end{aligned} \tag{6.19}$$

The ratio shows that the difference between the signal gain and the noise gain is the array gain. This intuitively makes sense because the electronic gain G_{a_m} affects both the signal and noise power, as previously mentioned. The component of the ratio in Equation 6.19 that contains the TRM and beamformer weighting coefficients (β_m, α_m) is an efficiency term that factors in the loss of array gain due to amplitude weighting. For the uniform weighting case where $\alpha_m = \beta_m = 1$, the efficiency term reduces to M, as shown in Equation 6.20.

$$\begin{aligned} \left. \frac{G_{signal_{AESA}}}{G_{noise_{AESA}}} \right|_{\alpha_m = \beta_m = 1} &= \left(D_e cos^{EF}(\theta) \cdot L_{array} \right) \cdot M \\ &= G_{array_{uniform\ weighting}}(\theta) . \end{aligned} \tag{6.20}$$

The ratio in Equation 6.19 also can be used to show how the taper loss is related to the array gain. This is shown in Equation 6.21,

$$\begin{aligned}\frac{G_{signal_{AESA}}}{G_{noise_{AESA}}} &= \left(D_e cos^{EF}(\theta) \cdot L_{array}\right) \frac{\left|\sum_{m=1}^{M} \alpha_m \beta_m\right|^2}{\sum_{m=1}^{M} \alpha_m^2 \beta_m^2} \\ &= \left(D_e cos^{EF}(\theta) \cdot L_{array}\right) \cdot M \cdot TL \\ &= G_{array}(\theta), \end{aligned} \tag{6.21}$$

where the taper loss, TL, is defined as

$$TL = \frac{1}{M} \cdot \frac{\left|\sum_{m=1}^{M} \alpha_m \beta_m\right|^2}{\sum_{m=1}^{M} \alpha_m^2 \beta_m^2}. \tag{6.22}$$

Using the definition of noise factor previously defined, the AESA cascaded noise factor can be written as

$$\begin{aligned} F_{AESA} &= \frac{\frac{S_{in}}{N_{in}}}{\frac{N_{out}}{N_{in}}} = \frac{S_{in}}{S_{out}} \cdot \frac{N_{out}}{N_{in}} \\ &= \frac{S_{in_m}}{G_{a_m} S_{in_m} \left|\sum_{m=1}^{M} \alpha_m \beta_m\right|^2} \cdot \frac{kTB \sum_{m=1}^{M} \alpha_m^2 \beta_m^2 G_{a_m} F_{TRM}}{kTB} \\ &= \frac{\sum_{m=1}^{M} \alpha_m^2 \beta_m^2 G_{a_m} F_{TRM}}{G_{a_m} \left|\sum_{m=1}^{M} \alpha_m \beta_m\right|^2} \\ &= \frac{\sum_{m=1}^{M} \alpha_m^2 \beta_m^2 \left(F_m + \frac{\left(\frac{1}{\beta_m^2} - 1\right)}{G_{a_m}}\right)}{\left|\sum_{m=1}^{M} \alpha_m \beta_m\right|^2} \\ &= \left(F_m - \frac{1}{G_{a_m}}\right) \frac{\sum_{m=1}^{M} \alpha_m^2 \beta_m^2}{\left|\sum_{m=1}^{M} \alpha_m \beta_m\right|^2} + \frac{\sum_{m=1}^{M} \frac{\alpha_m^2}{G_{a_m}}}{\left|\sum_{m=1}^{M} \alpha_m \beta_m\right|^2} \\ &= \left(\frac{\sum_{m=1}^{M} \alpha_m^2 \beta_m^2}{\left|\sum_{m=1}^{M} \alpha_m \beta_m\right|^2}\right) \left(F_m - \frac{1}{G_{a_m}} + \frac{\sum_{m=1}^{M} \alpha_m^2}{G_{a_m} \sum_{m=1}^{M} \alpha_m^2 \beta_m^2}\right) \\ &= \left(\frac{\sum_{m=1}^{M} \alpha_m^2 \beta_m^2}{\left|\sum_{m=1}^{M} \alpha_m \beta_m\right|^2}\right) \left(F_m + \frac{\frac{\sum_{m=1}^{M} \alpha_m^2}{\sum_{m=1}^{M} \alpha_m^2 \beta_m^2} - 1}{G_{a_m}}\right) \\ &= \left(\frac{1}{M \cdot TL}\right) \left(F_m + \frac{\frac{\sum_{m=1}^{M} \alpha_m^2}{\sum_{m=1}^{M} \alpha_m^2 \beta_m^2} - 1}{G_{a_m}}\right). \end{aligned} \tag{6.23}$$

Equation 6.23 shows that applying all of the weighting in the TRM attenuator degrades the noise factor compared to applying all of the weighting in the beamformer. For the case of all weighting in the beamformer ($\beta_m = 1$, for all M elements), F_{AESA} reduces to

$$F_{AESA_{beam\ former\ only}} = \left(\frac{\sum_{m=1}^{M} \alpha_m^2}{\left| \sum_{m=1}^{M} \alpha_m \right|^2} \right) (F_m) . \tag{6.24}$$

For the case of all weighting in the attenuators ($\alpha_m = \frac{1}{\sqrt{M}}$, for all M elements), F_{AESA} reduces to

$$F_{AESA_{attenuator\ only}} = \left(\frac{\sum_{m=1}^{M} \left(\frac{1}{\sqrt{M}} \right)^2 \beta_m^2}{\left| \sum_{m=1}^{M} \frac{1}{\sqrt{M}} \beta_m \right|^2} \right) \left(F_m + \frac{\frac{\sum_{m=1}^{M} \frac{1}{\sqrt{M}}^2}{\sum_{m=1}^{M} \frac{1}{\sqrt{M}}^2 \beta_m^2} - 1}{G_{a_m}} \right)$$

$$= \left(\frac{\sum_{m=1}^{M} \beta_m^2}{\left| \sum_{m=1}^{M} \beta_m \right|^2} \right) \left(F_m + \frac{\frac{M}{\sum_{m=1}^{M} \beta_m^2} - 1}{G_{a_m}} \right) . \tag{6.25}$$

The same taper loss expression applies for both Equations 6.24 and 6.25; however, the all attenuator weighting noise factor has an additional term that degrades the noise factor.

6.3.4 AESA *nth*-Order Intercept Point

In Chapter 4, it was shown that the *nth*-order intercept point can be expressed as

$$IP_n = \frac{1}{(1-n)} P_{O_n} - \frac{n}{(1-n)} P_{O_1} . \tag{6.26}$$

For an AESA a cumulative intercept point, $IP_{n_{AESA}}$, characterizing the linearity performance of the AESA can be derived based on Equation 6.26. In order to mitigate the impacts of third-order spurs and also second-order spurs for wideband systems ($f_{max} \geq 2f_{min}$), it is important to know $IP_{n_{AESA}}$ to minimize spur levels and enable linear operation. As will be shown in the next section, the spurious free dynamic range (SFDR) of the AESA is a function of $IP_{n_{AESA}}$.

To calculate $IP_{n_{AESA}}$, Equation 6.26 must be converted into into its scalar form, written as

$$ip_n = \frac{p_{o_n}^{\frac{1}{(1-n)}}}{p_{o_1}^{\frac{n}{(1-n)}}} . \tag{6.27}$$

The lower case letters for p and ip in Equation 6.27 represent voltage, and the upper case letters in Equation 6.26 represent power in dB such that $10log_{10}(p^r) = r10log_{10}(p) = rP$. From Equation 6.27, to calculate the intercept point for an AESA, expressions for both the linear output power and *nth*-order output power must be determined. Using the reference block diagram in Figure 6.12, the linear output power was shown to be

$$p_{o_{1_{AESA}}} = G_{a_m} S_{in_m} \left| \sum_{m=1}^{M} \alpha_m \beta_m \right|^2 . \tag{6.28}$$

To calculate $p_{o_{n_{AESA}}}$, the *nth*-order output power is calculated after the amplifier, and Equation 6.27 is rearranged for p_{o_n} as

$$p_{o_n} = p_{o_1}^n \, ip_n^{(1-n)} . \tag{6.29}$$

The *nth*-order output power after the amplifier in Figure 6.12 can then be expressed, using Equation 6.29, as

$$p_{o_{n_{TRM}}} = \left(G_{a_m} S_{in_m} \beta^2 \right)^n ip_{n_{TRM}}^{(1-n)} . \tag{6.30}$$

$p_{o_{n_{AESA}}}$ can then be written as

$$\begin{aligned} p_{o_{n_{AESA}}} &= \left| \sum_{m=1}^{M} \sqrt{\left(G_{a_m} S_{in_m} \beta^2 \right)^n ip_{n_{TRM}}^{(1-n)}} \; \alpha_m \right|^2 \\ &= \left(G_{a_m} S_{in_m} \right)^n \left| \sum_{m=1}^{M} \sqrt{ip_{n_{TRM}}^{(1-n)}} \; \alpha_m \beta_m^n \right|^2 . \end{aligned} \tag{6.31}$$

Equation 6.31 intuitively makes sense as it shows the the *nth*-order output power of the AESA is proportional to the linear output power raised to the power of n, similar to the general expression in Equation 6.29.

With expressions for both $p_{o_{1_{AESA}}}$ and $p_{o_{n_{AESA}}}$, the formula for the AESA intercept point is then

$$\begin{aligned} ip_{n_{AESA}} &= \frac{p_{o_{n_{AESA}}}^{\frac{1}{(1-n)}}}{p_{o_{1_{AESA}}}^{\frac{n}{(1-n)}}} \\ &= \frac{\left(\left(G_{a_m} S_{in_m} \right)^n \left| \sum_{m=1}^{M} \sqrt{ip_{n_{amplifier}}^{(1-n)}} \alpha_m \beta_m^n \right|^2 \right)^{\frac{1}{(1-n)}}}{\left(G_{a_m} S_{in_m} \left| \sum_{m=1}^{M} \alpha_m \beta_m \right|^2 \right)^{\frac{n}{(1-n)}}} \end{aligned}$$

$$= \frac{\left(\left|\sum_{m=1}^{M}\sqrt{ip_{n_{amplifier}}^{(1-n)}}\alpha_m\beta_m^n\right|^2\right)^{\frac{1}{(1-n)}}}{\left(\left|\sum_{m=1}^{M}\alpha_m\beta_m\right|^2\right)^{\frac{n}{(1-n)}}}$$

$$= \left(\frac{\left|\sum_{m=1}^{M}\sqrt{ip_{n_{amplifier}}^{(1-n)}}\alpha_m\beta_m^n\right|^2}{\left|\sum_{m=1}^{M}\alpha_m\beta_m\right|^{2n}}\right)^{\frac{1}{(1-n)}}. \tag{6.32}$$

If the AESA employs uniform weighting ($\alpha_m = \frac{1}{\sqrt{M}}$ and $\beta_m = 1$), then the AESA intercept point should be M multiplied by the TRM intercept point. To validate that this occurs for $ip_{n_{AESA}}$ in Equation 6.32, α_m is set to $\frac{1}{\sqrt{M}}$ and β_m is set to 1 with the resultant expression shown in Equation 6.33:

$$\begin{aligned}
ip_{n_{AESA}} &= \left(\frac{\left|\sum_{m=1}^{M}\sqrt{ip_{n_{amplifier}}^{(1-n)}}\alpha_m\beta_m^n\right|^2}{\left|\sum_{m=1}^{M}\alpha_m\beta_m\right|^{2n}}\right)^{\frac{1}{(1-n)}} \\
&= \left(\frac{\left|\sum_{m=1}^{M}\sqrt{ip_{n_{amplifier}}^{(1-n)}}\frac{1}{\sqrt{M}}\right|^2}{\left|\sum_{m=1}^{M}\frac{1}{\sqrt{M}}\right|^{2n}}\right)^{\frac{1}{(1-n)}} \\
&= \left(\frac{ip_{n_{amplifier}}^{(1-n)}M}{M^n}\right)^{\frac{1}{(1-n)}} \\
&= M\, ip_{n_{amplifier}}.
\end{aligned} \tag{6.33}$$

This same result is derived in [Holzman, 1996].

6.3.5 AESA Spurious Free Dynamic Range

A final parameter that is important for AESAs is SFDR. This defines the ability of the AESA to amplify incoming signals above the noise floor. The maximum signal is the linear output value when the *nth*-order output power is equal to the noise floor power. The minimum signal is the smallest linear output power that is above the noise floor. For the purposes of this derivation it is assumed that the minimum linear output power equals the output noise power. In a system, the minimum linear signal power must be at a minimum threshold value that is greater than the noise floor power in order to increase the probability of detection. Additionally, the *nth*-order output power is typically specified to be at some value less than the output noise power to ensure minimal spurious products will be amplified above the noise floor for a potential false detection.

Mathematically the SFDR is defined as

$$SFDR = P_{o_{1_{max}}} - P_{o_{1_{min}}}. \tag{6.34}$$

$P_{o_{1_{min}}}$ is set to N_{out} as previously described. $P_{o_{1_{max}}}$ can be expressed using the equation for the *nth*-order output power, where

$$\begin{aligned} P_{o_n} &= nP_{o_1} + (1-n)IP_n \\ &= nP_{o_{1_{max}}} + (1-n)IP_n. \end{aligned} \tag{6.35}$$

Setting P_{o_n} equal to N_{out} and rearranging Equation 6.35 in terms of $P_{o_{1_{max}}}$ gives the following expression

$$P_{o_{1_{max}}} = \left(\frac{n-1}{n}\right) IP_n + \left(\frac{1}{n}\right) N_{out}. \tag{6.36}$$

Substituting Equation 6.36 into Equation 6.34 with $P_{o_{1_{min}}} = N_{out}$ gives

$$\begin{aligned} SFDR &= \left(\frac{n-1}{n}\right) IP_n + \left(\frac{1}{n}\right) N_{out} - N_{out} \\ &= \left(\frac{n-1}{n}\right) \left(IP_n - N_{out}\right). \end{aligned} \tag{6.37}$$

Intuitively, this makes sense. By increasing IP_n, the dynamic range of the AESA will be increased allowing larger signals at the output of the AESA. By decreasing or minimizing the output noise power, smaller signals will be able to be seen above the noise floor. Finally, with the previous derivation for $IP_{n_{AESA}}$, the SFDR of the AESA can be expressed as

$$\begin{aligned} SFDR_{AESA} = \left(\frac{n-1}{n}\right) \Bigg[& \left(\frac{\left| \sum_{m=1}^{M} \sqrt{ip_{n_{amplifier}}^{(1-n)}} \alpha_m \beta_m^n \right|^2}{\left| \sum_{m=1}^{M} \alpha_m \beta_m \right|^{2n}} \right)^{\frac{1}{(1-n)}} \\ & - kTB \sum_{m=1}^{M} \alpha_m^2 \beta_m^2 G_{a_m} \left(F_m + \frac{\left(\frac{1}{\beta_m^2} - 1\right)}{G_{a_m}} \right) \Bigg]. \end{aligned} \tag{6.38}$$

References

Holzman, E.H. "Intercept points of active phased array antennas." *IEEE MTT-S Digest*, pp. 999–1002, 1996.

Pettai, R. *Noise in Receiving Systems*. John Wiley & Sons, 1984.

7

AESA Architectures

Key Concepts

- Baseline Architecture
- Subarray Architecture
- Overlapped Subarray Architecture
- Elemental DBF Architecture
- Adaptive Beamforming

7.1 Introduction

In Chapter 2, the basic fundamentals of AESA theory were discussed. Important covered topics germane to AESA design were beamwidth, instantaneous bandwidth (IBW), grating lobes, and so forth. This chapter builds upon these fundamentals.

7.2 Baseline Architecture

The baseline architecture for an AESA assumes phase shifter steering at each AESA element with an analog beamformer used to coherently add the signal from all the elements. This AESA architecture topology is illustrated in Figure 7.1. For review, several features of an AESA's scan behavior as illustrated in Figure 7.2 are:

- The overall AESA pattern can be calculated using pattern multiplication, which is the product of a single array element pattern and the *AF*. For AESAs whose

Active Electronically Scanned Arrays: Fundamentals and Applications, First Edition. Arik D. Brown.

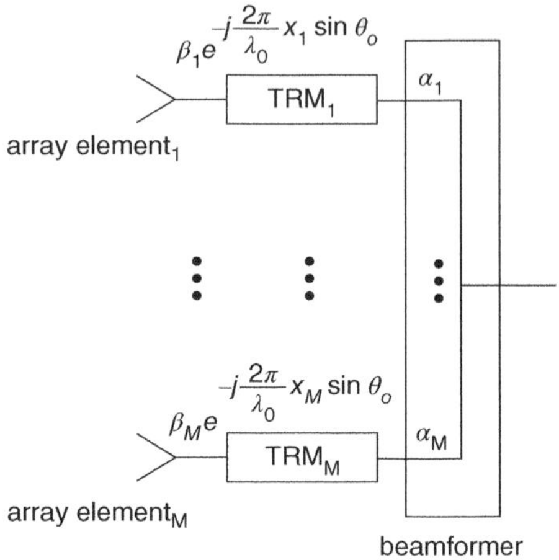

Figure 7.1 Baseline AESA architecture has an analog beamformer with phased delay at each element.

array element patterns can not be represented with only a single pattern, the element pattern for each element must be included.

- The AESA beamwidth increases with scan. This is different from an MSA, which only has a boresight pattern that is mechanically articulated across the FOV.
- The element pattern of the array elements causes a loss in the overall AESA pattern due to the element pattern roll-off. This loss is referred to as scan loss and is a driver of the AESA array size. The effective area must be made large enough to accommodate the loss at the maximum scan angle.
- The *AF* is a function of frequency, array element spacing, angle, and scan angle.
- For a uniform rectangular distribution, the 1st SL is 13 dB down from the main beam. The SLs can be lowered by applying an amplitude taper across the array, which significantly reduces the SLL. Figure 7.2 shows an AESA with a 35 dB, nbar = 5, and Taylor weighting. Using weighting causes the array to suffer an additional loss in gain, which is referred to as taper loss.
- AESAs in real applications have phase and amplitude errors distributed over all the elements in the AESA. These errors degrade the SL structure and increase peak and average SLLs.

The baseline AESA architecture shown in Figure 7.1 has some limitations. Using phase shifters, the AESA suffers from beam squint for large IBWs. Time delay steering can be employed but is lossy and prohibitive for large numbers of array elements. Additionally, the baseline architecture provides no opportunity for scalability. To build an AESA that is even a fraction larger in aperture area requires a completely new AESA to be built. There are no building blocks that can be used

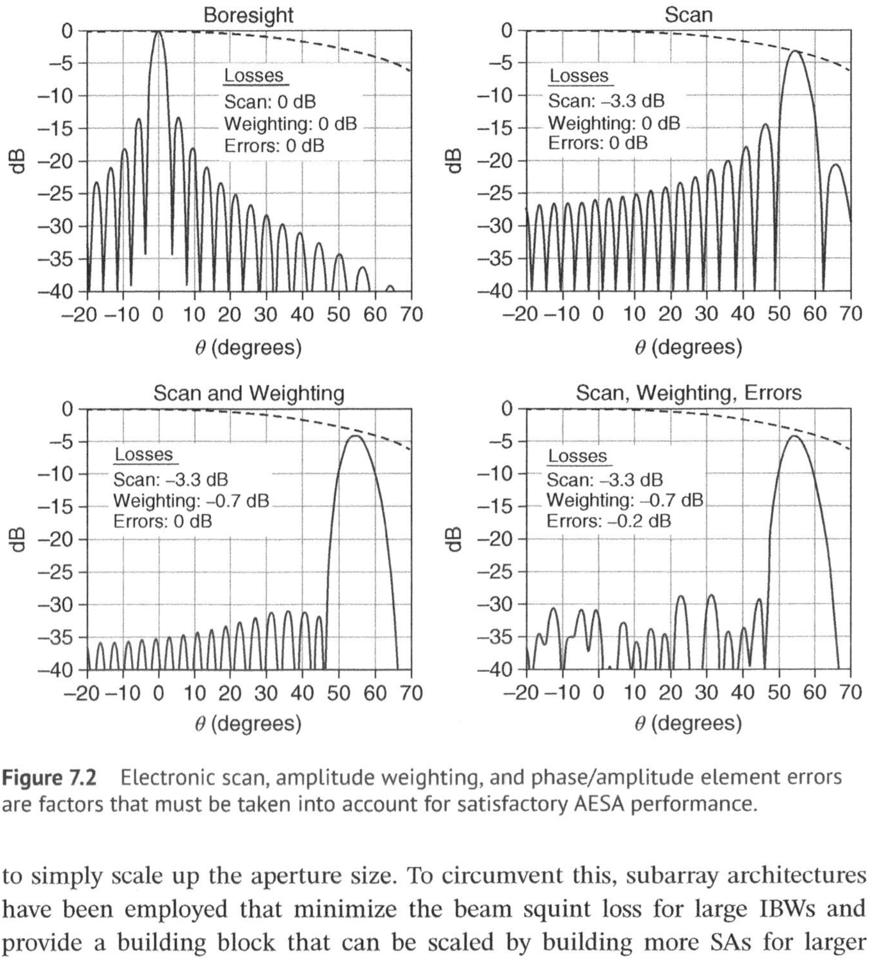

Figure 7.2 Electronic scan, amplitude weighting, and phase/amplitude element errors are factors that must be taken into account for satisfactory AESA performance.

to simply scale up the aperture size. To circumvent this, subarray architectures have been employed that minimize the beam squint loss for large IBWs and provide a building block that can be scaled by building more SAs for larger aperture sizes. This takes advantage of the efficiencies (economies of scale) gained in manufacturing when building large quantities of the same item. Subarray architectures will be covered later in this chapter. Overlapped subarray architectures will also be discussed, which is an improvement to the baseline subarray architecture for minimizing off-tune grating lobes.

Ideally, the ultimate AESA would not have TRMs or a beamformer. Instead, an HPA, LNA, and DAC/ADC would be placed at each element in the AESA. The correct delay and amplitude would be implemented digitally. This enables a better NF for sensitivity, less loss for transmit power, and a size reduction by eliminating the need for TRMs and a beamformer. Additionally, there would be no IBW limitation since digital time delay would be used. For AESAs that are on the order of thousands of elements this does have some issues for implementation. The power required to implement a DAC/ADC at each element is daunting, in addition to the throughput required for digitally distributing thousands of channels and applying beamforming for one or more beams. Technology is currently advancing where this is becoming closer to a reality as evidenced by DARPA's Arrays at Commercial Timescales (ACT) program. Even with such advances, the ultimate elemental DBF AESA for large numbers of elements is still in the future. Comparisons between the baseline architecture and an elemental DBF architecture will be shown in this chapter.

7.3 Subarray Architectures

In Chapter 2, the topic of IBW was covered in detail. It was shown that the IBW is inversely proportional to the length of an aperture. For most current and emerging systems that require AESAs, larger IBWs are desired on the order of hundreds of MHz. For a given aperture size, the IBW is fixed. If the IBW is less than that desired, then the AESA's performance will suffer from beam squint. This is depicted in Figure 7.3. Additionally, for applications where greater sensitivity is required via larger AESA apertures (on the order of 1 m^2 or larger), IBW is limited. Larger apertures will inherently have limited IBW.

In order to provide wide IBWs, true time delay can be used at each element in place of the phase shifters shown in Figure 7.1. However, time delay at each element would be expensive, lossy, and contain an additional source of error (Skolnik, 1990). The practical solution is to use subarrays, which are multiple single array elements grouped together. The subarrays form the entire array, and the subarrays can be thought of as having the effective element pattern that is used to compute the entire pattern. A subarray architecture topology is shown in Figure 7.4. The analog beamformer in Figure 7.4 performs the combining on

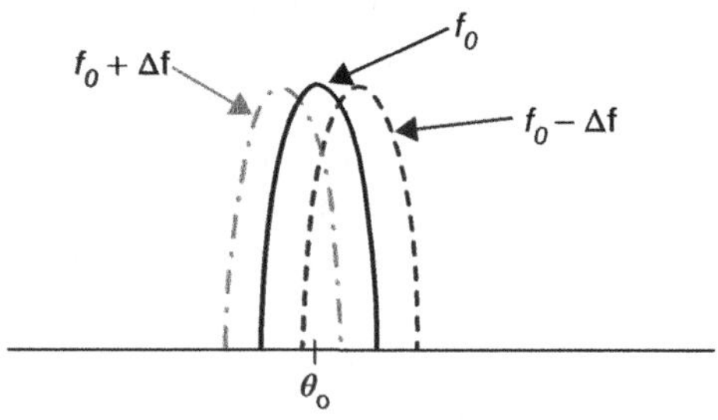

Figure 7.3 SA architectures have their foundation in the pursuit of wider IBW AESAs. The baseline architecture suffers from reduced IBW due to phase shifter implementation. SA architectures attempt to overcome this limitation.

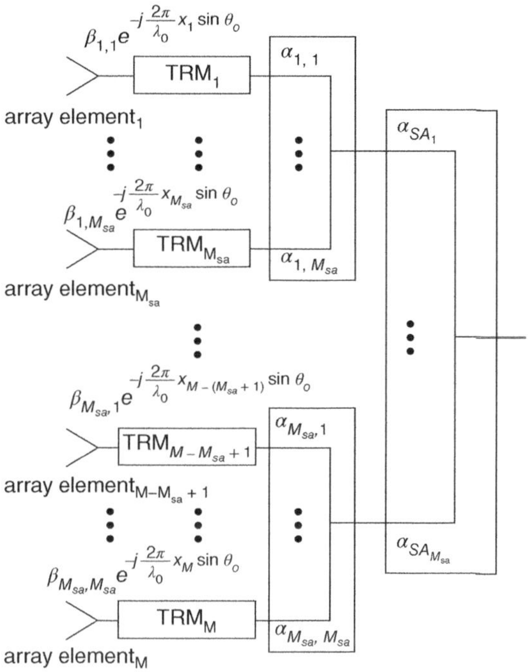

Figure 7.4 SA beamforming consists of dividing the AESA into SAs and using two levels of beamforming to generate a beam. The 1st level of beamforming occurs at the SA level, and the 2nd level of beamforming occurs at the backend of the AESA.

a subarray-to-subarray level in contrast to the element-to-element combining illustrated in Figure 7.1.

In addition to alleviating the issues previously discussed in this section related to IBW, subarray architectures can also reduce the number of control elements required in an AESA. For applications that have limited scan requirements, phase shifters can be employed behind groups of subarrays as opposed to each element. However, this provides very limited scan performance, as will be discussed later in this chapter. Another benefit subarrays provide is a larger building block for manufacturing and fabrication. Instead of building many individual modules with a single phase shifter, subarray building blocks can be built, which is quite attractive for larger arrays.

Adaptive beamforming is a technique that leverages subarray and partial elemental DBF architectures. In some AESAs, DBF can be implemented in one dimension with analog beamforming in the other dimension. Both of these approaches enable the ability to apply adaptive weighting on receive to counter intentional jammers. This is commonly referred to as electronic counter-countermeasures (ECCM). The basic theory will be discussed later in this chapter with pattern examples. A key feature of adaptive beamforming is enhancing the target signal while cancelling the disturbance (Skolnik, 1990).

7.4 Subarray Pattern Formulation

This section will focus on the 1D formulation to explain and describe AESA pattern behavior for subarray AESA architectures. The 2D formulation for subarrays is left as an exercise for the reader but can be readily derived as outlined by the formulation shown in Chapter 2 for non-subarrayed AESAs. The 1D formulation is more than satisfactory to provide the insight into subarray AESA pattern performance.

In Chapter 2, the pattern for a 1D AESA was shown to be

$$F(\theta) = cos^{\frac{EF}{2}}\theta \cdot \sum_{m=1}^{M} a_m e^{j\left(\frac{2\pi}{\lambda}x_m sin\,\theta - \frac{2\pi}{\lambda_o}x_m sin\,\theta_o\right)}. \tag{7.1}$$

For a subarrayed architecture, the pattern formulation is similar to Equation 7.1 and can be expressed in the same manner. We begin by assuming an AESA composed of M individual elements. These M elements are divided into P subarrays, where the number of elements per subarray, R, is $\frac{M}{P}$. The number of elements in the array does not have to be an integer multiple of the subarray elements. In the case of overlapped subarrays, the subarrays on the edge of the array may have fewer elements per subarray.

Using Equation 7.1, the pattern for the subarrays can be written as

$$F_{SA}(\theta) = cos^{\frac{EF}{2}}\theta \cdot \sum_{r=1}^{R} a_r e^{j\left(\frac{2\pi}{\lambda}x_r sin\theta - \frac{2\pi}{\lambda_o}x_r sin\theta_o\right)}. \tag{7.2}$$

In order to express the complete pattern, a term needs to be added to Equation 7.2 for the analog beamforming. This results in the following expression

$$F(\theta) = F_{SA}(\theta) \cdot \sum_{p=1}^{P} b_p e^{j\left(\frac{2\pi}{\lambda}x_p sin\theta - \frac{2\pi}{\lambda_o}x_p sin\theta_o\right)}. \tag{7.3}$$

Combining Equations 7.2 and 7.3, and substituting AF_{SA} for the summation in Equation 7.2, the expression for a subarrayed AESA pattern is the following

$$F(\theta) = EP \cdot \sum_{p=1}^{P} b_p \cdot AF_{SA_p} e^{j\left(\frac{2\pi}{\lambda}x_p sin\theta - \frac{2\pi}{\lambda_o}x_p sin\theta_o\right)}. \tag{7.4}$$

If we assume that all the subarrays have the same AF, Equation 7.4 can be further simplified as

$$F(\theta) = \left(EP \cdot AF_{SA_p}\right) \cdot \sum_{p=1}^{P} b_p \cdot e^{j\left(\frac{2\pi}{\lambda}x_p sin\theta - \frac{2\pi}{\lambda_o}x_p sin\theta_o\right)}. \tag{7.5}$$

Equation 7.5 closely resembles Equation 7.1. The difference is that the effective element pattern is the element pattern of a single element multiplied by the *AF*. Additionally, the summation is done over the number of subarrays instead of the number of elements. Simplifying Equation 7.5 more we arrive at the following expression

$$F(\theta) = (EP \cdot AF_{SA_p}) \cdot AF_p = EP_{SA} \cdot AF_p. \tag{7.6}$$

Equation 7.6 is a very intuitive result. It shows that the overall pattern for a subarrayed AESA reduces to pattern multiplication of the subarray element pattern with the AF of the backend beamforming. This is very useful in understanding the implications of using true time delay or digital beamforming across the subarrays, instead of using phase shifters, as shown in Equation 7.5. These implications will be elaborated on in the following section.

7.5 Subarray Beamforming

There are multiple AESA architecture topologies that can be used for implementing subarray beamforming. These involve using one or multiple combinations of phase shifter, time delay, and digital beamforming. In this chapter, we will discuss three different architecture approaches with the limitations and advantages that each one provides. These architectures are illustrated in Figure 7.5 and will be expounded upon in the following subsections.

7.5.1 Subarray Phase Shifter Beamforming

Figure 7.6 shows an AESA subarray architecture employing only phase shifters at the subarray level. As mentioned earlier in this chapter, this is advantageous because the number of control elements required can be dramatically reduced. Instead of using a phase shifter at each element to electronically scan the beam, phase shifters are used only at the subarray level. Expanding Equation 7.5, the equation for a subarrayed pattern is

$$F(\theta) = \left(EP \cdot \sum_{r=1}^{R} a_r e^{j\left(\frac{2\pi}{\lambda}x_r sin\theta - \frac{2\pi}{\lambda_o}x_r sin\theta_o\right)} \right) \cdot \sum_{p=1}^{P} b_p \cdot e^{j\left(\frac{2\pi}{\lambda}x_p sin\theta - \frac{2\pi}{\lambda_o}x_p sin\theta_o\right)}. \tag{7.7}$$

For the array in Figure 7.6, the array factor term in Equation 7.7 must be modified because there are no phase shifters on the element level. Equation 7.7 then becomes

$$F(\theta) = \left(EP \cdot \sum_{r=1}^{R} e^{j\left(\frac{2\pi}{\lambda}x_r sin\theta\right)} \right) \cdot \sum_{p=1}^{P} b_p \cdot e^{j\left(\frac{2\pi}{\lambda}x_p sin\theta - \frac{2\pi}{\lambda_o}x_p sin\theta_o\right)}. \tag{7.8}$$

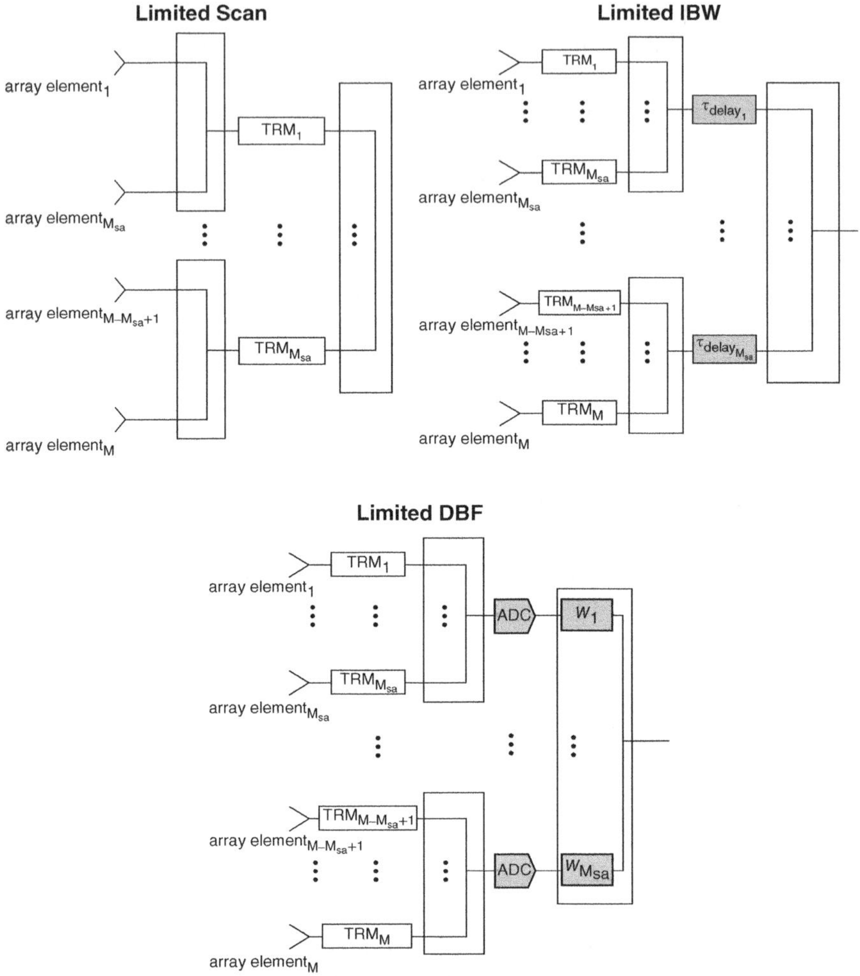

Figure 7.5 Various types of SA architectures.

Figure 7.7 illustrates the expression in Equation 7.8. Although the number of control elements has been reduced, the scan capability has also been reduced. Because the AF of the subarray does not steer with the beam, unwanted high sidelobes are the result in the pattern. This is shown in Figure 7.8. Phase shifter steering only at the subarray level is then only viable for applications where the scan requirement is limited.

It is important to point out that even with time delay at the subarray level instead of phase delay, the results are the same. With no elemental control, the

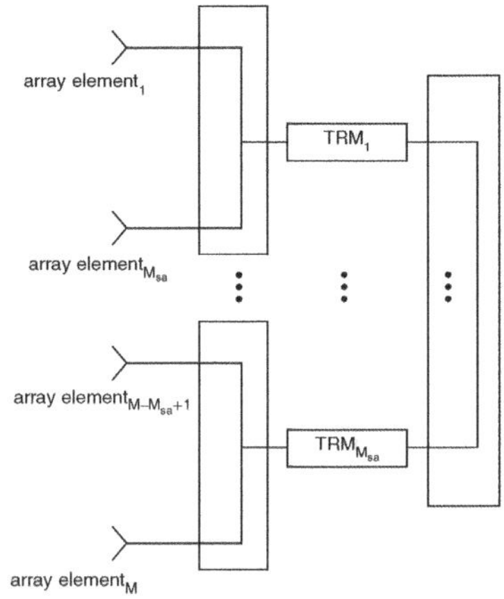

Figure 7.6 This SA architecture suffers from limited scan because there is no elemental scanning control.

AF of the subarray cannot be electronically steered and thus limits the AESA's ability to scan.

7.5.2 Subarray Time Delay Beamforming

The previous section showed that in order to not restrict the scan capability of a subarrayed AESA, some type of elemental control is required. This allows the subarray *AF* to scan with the main beam. Phase shifter delay or time delay can be employed on the element level. However, elemental time delay adds undesired complexity especially for large AESAs (100s or 1000s of elements). This leads to the approach shown in Figure 7.9. The expression for the full array pattern using Equation 7.7 is

$$F(\theta) = \left(EP \cdot \sum_{r=1}^{R} a_r e^{j\left(\frac{2\pi}{\lambda} x_r sin\theta - \frac{2\pi}{\lambda_0} x_r sin\theta_o\right)} \right) \cdot \sum_{p=1}^{P} b_p \cdot e^{j\frac{2\pi}{\lambda} x_p (sin\theta - sin\theta_o)}. \quad (7.9)$$

Using phase delay at the element level and time delay at the subarray level reduces the number of time delay devices required and provides excellent scan performance. Implementing control on the element level allows the AF of the subarray to be scanned with the array. Figure 7.10 shows an example of scanning the AESA with the approach shown in Figure 7.9. The array is scanned to 30° at the tune frequency, f_o, and the pattern is well behaved, as expected.

A drawback of this approach is the limited IBW that results. Looking at Equation 7.9, the *AF* of the subarray (in parentheses) has a maximum value at the tune frequency (i.e., $\lambda = \lambda_o$); however, at frequencies offset from the tune

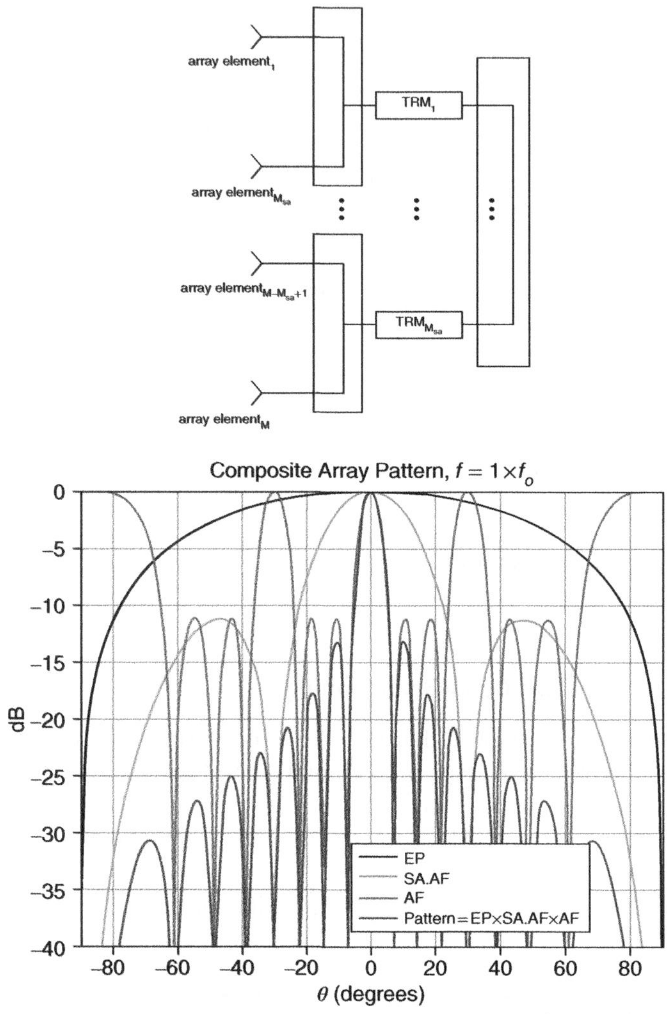

Figure 7.7 At boresite the limited scan SA architecture performs well.

frequency, the AF squints and does not have a maximum value at the off-tune frequency (i.e., $\lambda \neq \lambda_o$). This is illustrated in Figure 7.11. This limitation in IBW is still better than a non-subarrayed AESA. In Chapter 2, the IBW of an AESA was shown to be

$$IBW = \frac{c}{L sin\theta_o}. \tag{7.10}$$

For a subarrayed AESA with time delay at the subarray level, the IBW is limited by the size of the subarray and not the array and can be expressed as

$$IBW = \frac{P \cdot c}{L sin\theta_o}, \tag{7.11}$$

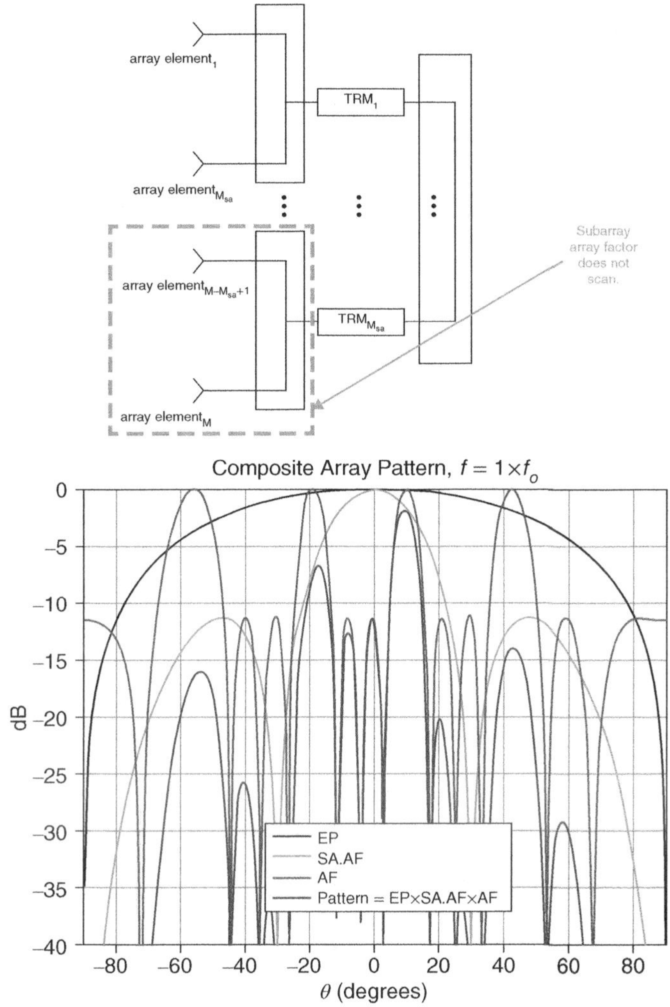

Figure 7.8 Without the ability to scan the SA, the overall scan capability of the AESA is limited.

where P is the number of subarrays. Comparing Equations 7.10 and 7.11, the IBW is increased by the number of subarrays, P. This is because for a fixed aperture lenght L, more SAs means the SA size decreases for increased IBW. However, with the increased IBW there is an accompanying gain loss (no free lunch!), which is described in (Skolnik, 1990) as

$$loss\ in\ gain \approx 1 - \left(\frac{sin\left[\left(\frac{\pi}{4}\right) sin\theta_o\right]}{(\frac{\pi}{4}) sin\theta_o} \right)^2 . \tag{7.12}$$

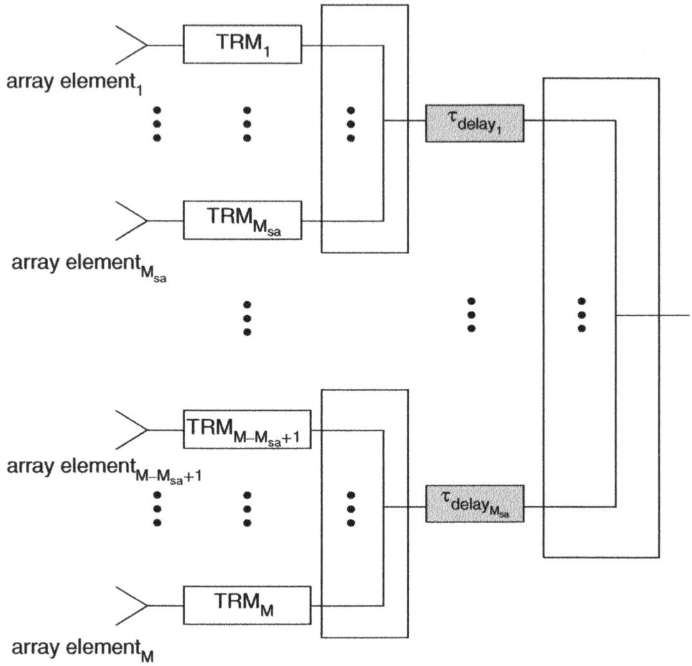

Figure 7.9 Adding phase shifters to the SA elements via the TRMs enables the SA to scan removing the scan limitation of the limited scan SA architecture.

This decrease in gain is due to the increased beamwidth of the SA. As the SA beamwidth increases, the SLs from the backend *AF* increase as well.

7.5.3 Subarray Digital Beamforming

Implementing phase delay at the element level and time delay at the subarray level provides robust scan performance with limitations on IBW. These limitations still provide an advantage over a non-subarrayed AESA of the same size that has phase-only steering. An alternative to time delay steering at the subarray level is to place a receiver channel at each subarray and combine the subarrays digitally. This is referred to as digital beamforming and is shown in Figure 7.12.

Mathematically, the pattern formulation is similar to that shown in Equation 7.9. Digital beamforming (DBF) is an enabler for generating multiple simultaneous AESA beams with full aperture gain. By digitally multiplexing the output signal from the ADCs, multiple simultaneous beams can be created digitally. By adjusting the digital weights, multiple beams scanned in different directions can be created simultaneously. The electronic scan for these beams is limited to the beamwidth of the SA. This is extremely beneficial because each of these beams has full aperture gain. This is illustrated in Figure 7.13. The same

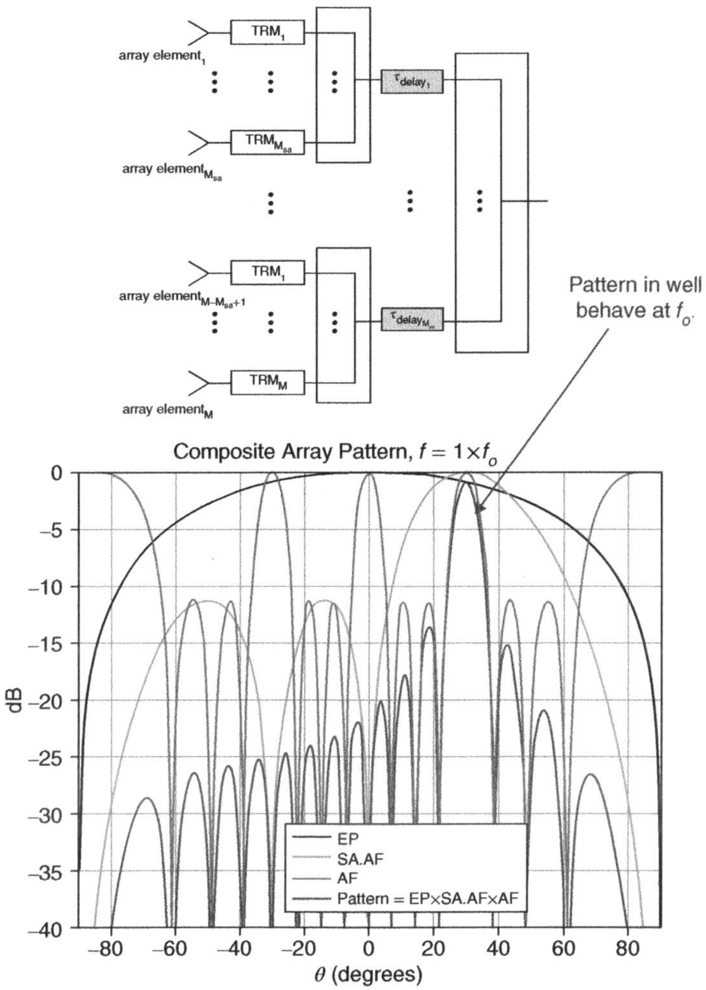

Figure 7.10 At the commanded tune frequency, the SA architecture pattern is well behaved.

issues with IBW exist with DBF. In order to minimize this effect, the subarray aperture length is designed to match the IBW required.

7.6 Overlapped Subarrays

For both subarray time delay beamforming and subarray digital beamforming, the IBW is limited by the shape of the beam pattern of the subarray. As the frequency is varied from the operational center frequency, the subarray pattern squints and

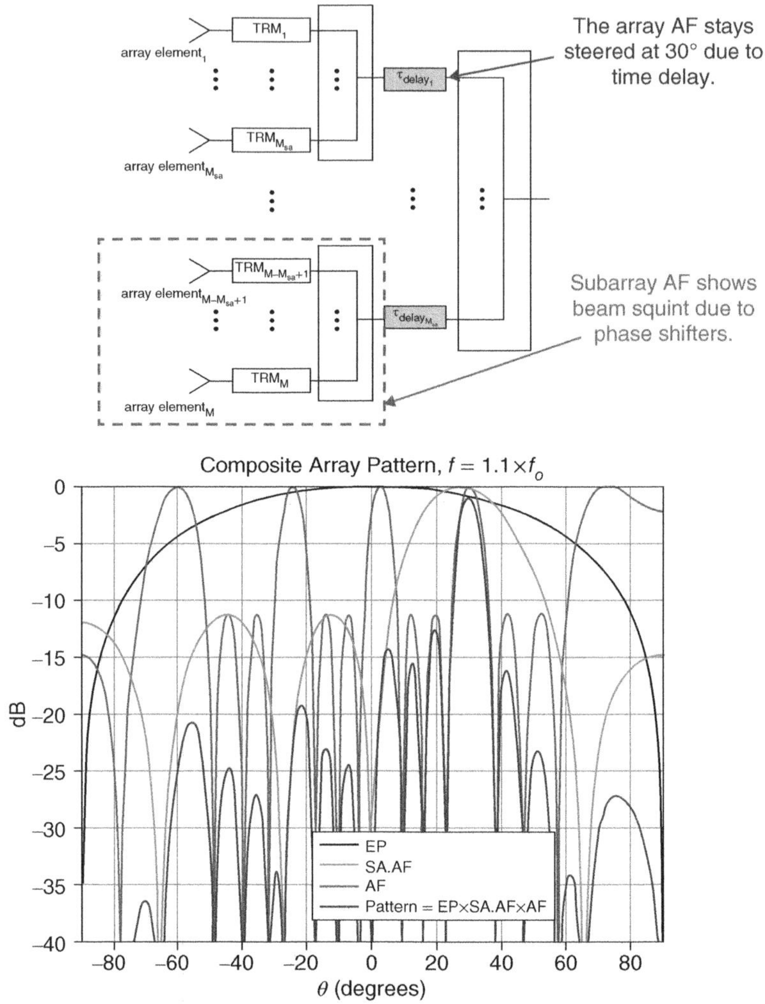

Figure 7.11 The IBW of this SA architecture is limited by the size of the SA pattern at off-tune frequencies. The backend *AF* scales with the *AF* SA pattern, leading to increased SLLs.

allows the grating lobes from the backend AF to appear in the pattern. In order to mitigate this effect, a subarray pattern that provides a windowing effect similar to a filter is required. This can be accomplished using overlapped subarrays. Figure 7.14 shows an example of an overlapped subarray AESA architecture. Adjacent subarrays are used to form an overlapped subarray AF. This array factor provides the windowing effect to reduce grating lobes at off-tune frequencies.

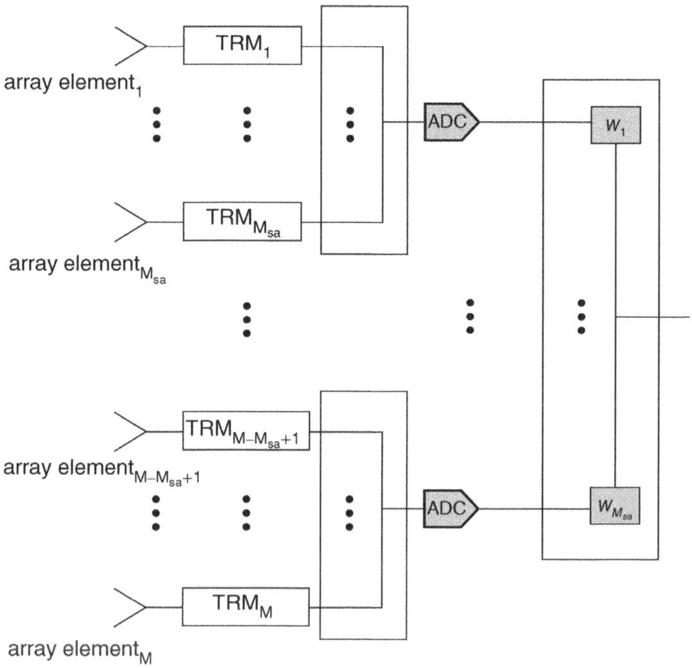

Figure 7.12 Replacing the time delay behind each SA with an ADC enables the ability to digitally create simultaneous beams.

Ideally, a $\frac{sinx}{x}$ distribution would provide a window-like spatial pattern distribution, as shown in Figure 7.15.

The overlapped subarray AF can be expressed as

$$\begin{aligned} AF_{OSA}(\theta) = \; & \sum_{r=1}^{R} A_{r,p} e^{j\left(\frac{2\pi}{\lambda} x_{r,p} sin\theta - \frac{2\pi}{\lambda_0} x_{r,p} sin\theta_o\right)} \\ & + \sum_{p=1}^{R} B_{r,p+1} e^{j\left(\frac{2\pi}{\lambda} x_{r,p+1} sin\theta - \frac{2\pi}{\lambda_0} x_{r,p+1} sin\theta_o\right)}. \end{aligned} \tag{7.13}$$

Equation 7.13 assumes a 2:1 overlap of adjacent subarrays. The total expression for the overlapped subarray AESA can be written as

$$\begin{aligned} F(\theta) = \; & \left(EP \cdot \sum_{r=1}^{R} A_{r,p} e^{j\left(\frac{2\pi}{\lambda} x_{r,p} sin\theta - \frac{2\pi}{\lambda_0} x_{r,p} sin\theta_o\right)} \right. \\ & \left. + \sum_{p=1}^{R} B_{r,p+1} e^{j\left(\frac{2\pi}{\lambda} x_{r,p+1} sin\theta - \frac{2\pi}{\lambda_0} x_{r,p+1} sin\theta_o\right)} \right) \\ & \cdot \sum_{p=1}^{P-1} b_p \cdot e^{j\frac{2\pi}{\lambda} x_p (sin\theta - sin\theta_o)}. \end{aligned} \tag{7.14}$$

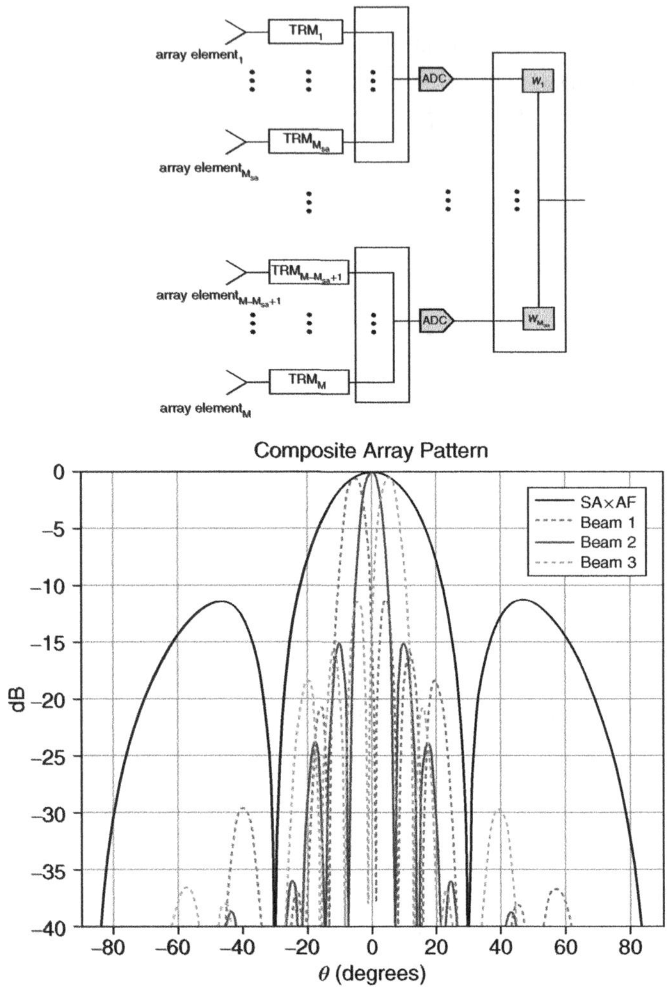

Figure 7.13 Multiple simultaneous beams are illustrated using the SA DBF architecture. The beams are only limited by the width of the SA pattern. This same performance can be achieved wherever the SA pattern is scanned.

Equation 7.14 is similar in form to Equation 7.9. The only difference is that the overlapped subarray *AF* replaces that of the non-overlapped subarray *AF*. The subarrays on the end of the array in Figure 7.14 have a contribution, as they form half of an overlapped subarray without a matching half. These are left out for simplicity. Figures 7.16 and 7.17 illustrate the benefit of overlapped subarrays.

The non-overlapped subarray at off-tune frequencies shows grating lobes whose origin is the *AF* of the backend beamformer. The overlapped subarray pattern, in

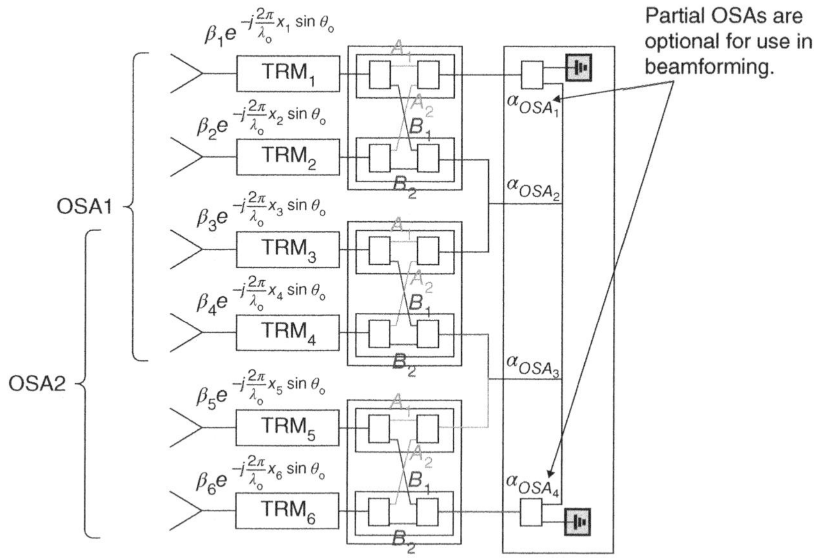

Figure 7.14 The overlapped SA architecture is used to decrease the size of the SA *AF* beamwidth for improved IBW performance by limiting grating lobes from the backed *AF* SLs.

contrast, performs very well away from the tune frequency. As previously mentioned, the *AF* of the overlapped subarray spatially attenuates the *AF* of the back-end beamformer and provides superior IBW performance in comparison to the non-overlapped subarray pattern.

7.7 Elemental DBF Architecture

As previously discussed, a fully elemental DBF (EDBF) architecture does not have TRMs or a beamformer. Instead, each array element has its own HPA, LNA, and DAC/ADC. This is depicted in Figure 7.18. Instead of the amplitude and delay weighting being implemented via analog, in this architecture, this is done digitally. Additionally, with sampled transmit and receive signals from the DAC and ADC, an analog beamformer is not required either.

For applications requiring a large IBW, EDBF provides incredible benefit. As shown in Chapter 2, when electronically steering the main beam with true time delay (TTD), there is no beam squint loss. The AESA can scan its beam over the FOV, and off-tune frequencies will not incur squint loss. This is illustrated in Figure 7.19. The analog AESA with phase shifter steering incurs a loss at frequencies that are not equal to the tune frequency. The EDBF architecture,

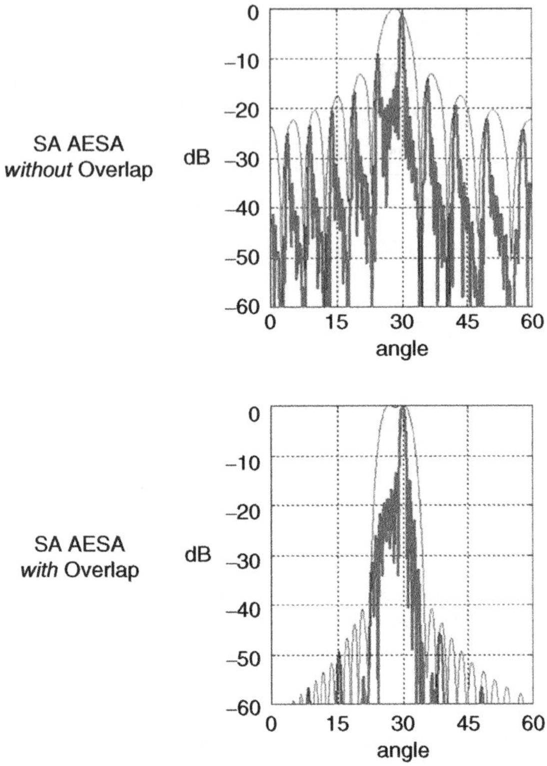

Figure 7.15 Overlapped SA approach is to decrease the SA AF beamwidth and limit its SLL.

however, does not. This is extremely useful for large AESAs with IBWs that are 500 MHz or greater.

Finally, another benefit of EDBF AESAs is the ability to create multiple simultaneous beams on transmit and receive without multiple analog beamformers. Multiple beams can be generated with a subarray architecture that is digitized at the subarray level, but they are still limited by the IBW of the SA. Figure 7.20 shows multiple beams simultaneously scanned across the FOV with no limitation due to a fixed number of SAs.

Low-frequency AESAs are a candidate for EDBF because their number of elements is small, on the order of 100. This makes EDBF for low frequency a viable approach with the current state of technology. Managing the power required for 100s of DAC/ADCs is still a challenge but in the realm of possible with current technology. For higher frequencies where the number of elements grows from 100s to 1000s, EDBF is still very difficult from a power and data rate/throughput standpoint (Melvin and Scheer, 2013).

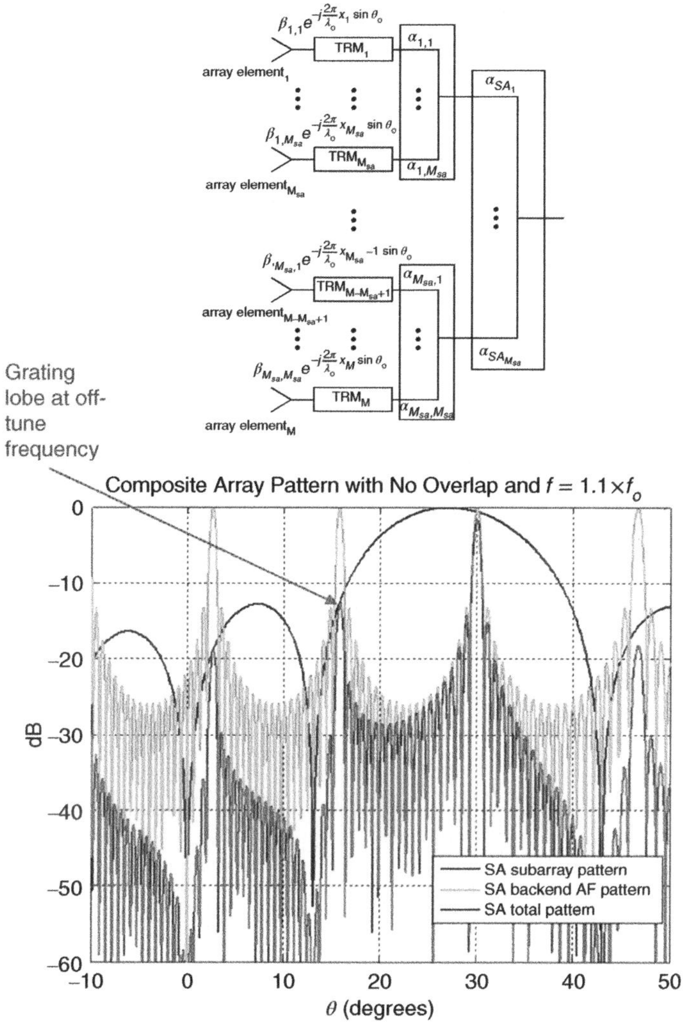

Figure 7.16 For the non-overlapped SA architecture, grating lobes due to the backend *AF* result in high SLs for the full array pattern.

7.8 Adaptive Beamforming

As previously discussed, SL weighting can be used to negate the effects of jamming signals that enter the SLs of the AESA. However, as shown in Chapters 2 and 5, there is an associated taper loss that reduces the sensitivity of the beam. The larger the taper for decreased SLLs, the greater the loss in main beam antenna gain.

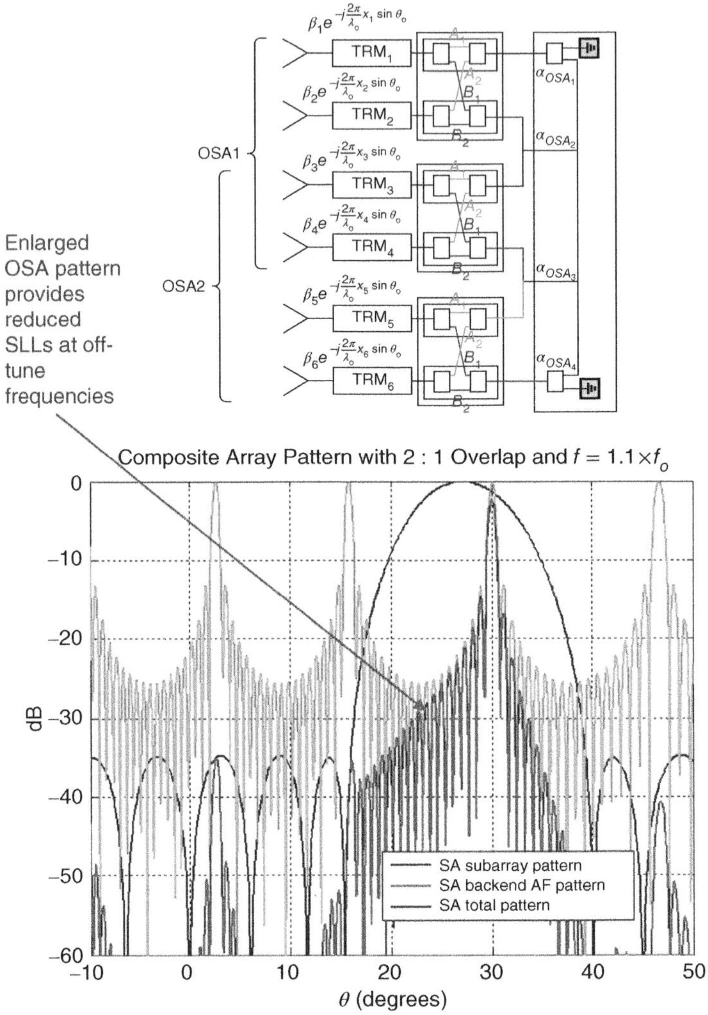

Figure 7.17 Using overlapped SAs the SA *AF* beamwidth is made narrower while weighting its SLs to minimize the effects of the backend AF. This results in excellent performance at off-tune frequencies.

Table 7.1 shows the sensitivity loss for increasing Taylor weighting as compared to the uniform weighting case.

As was mentioned in Chapter 2, even with amplitude weighting, phase and amplitude errors distributed across the AESA elements put a fundamental limit on how low the SLLs can be made.

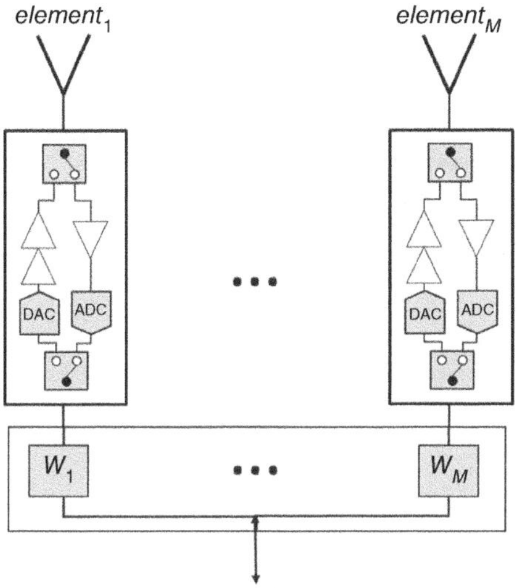

Figure 7.18 EDBF completely eliminates analog beamforming. All beamforming is done digitally, with no IBW or SA limitations.

Table 7.1 Decreasing the SLL to combat ECM also increases the loss in gain. Adaptive beamforming is a technique that can provide ECCM without prohibitively low SLLs or unacceptable sensitivity loss.

2D Taylor Weighting, 1600 Elements (40× 40)		
SLL	**nbar**	**Loss**
−20 dB	5	−0.3 dB
−25 dB	5	−0.8 dB
−30 dB	5	−1.3 dB
−35 dB	5	−1.8 dB
−40 dB	5	−2.3 dB

An alternative to employing large amplitude weighting is adaptive beamforming (ABF). ABF works by automatically adjusting the beamformer weights to minimize jammer power that is incident in the SLs (Melvin and Scheer, 2013). Figure 7.21 shows an M-element AESA with an ADC at each element. For simplicity the receive case is shown without including the LNA. This depicts an elemental approach, but the same SA theory applies with the SA pattern being used for the analysis in place of the array element pattern.

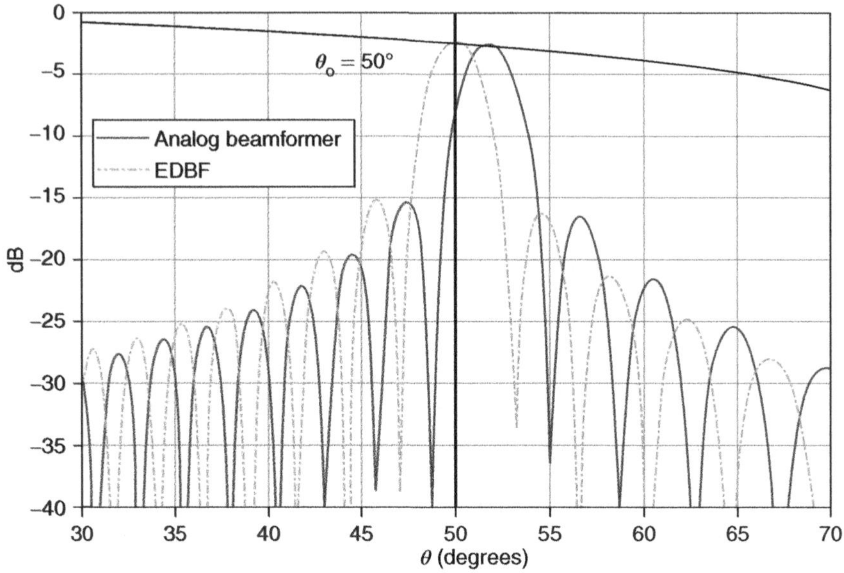

Figure 7.19 Beam squint is eliminated with EDBF vs. analog beamforming. At the commanded scan angle of 50° in the figure, phase shifter steering causes severe squint but is not an issue for EDBF.

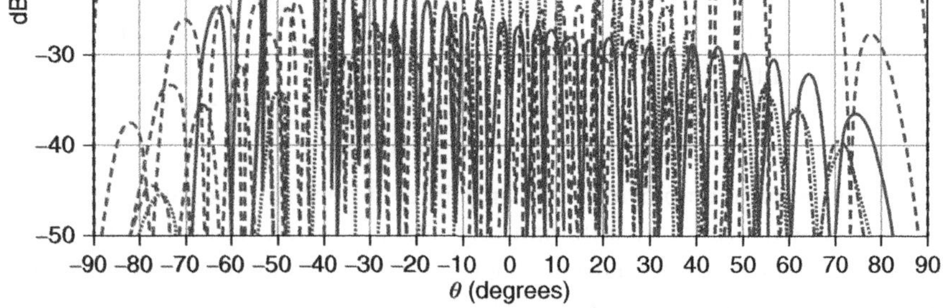

Figure 7.20 Multiple simultaneous beams can be formed with EDBF without being limited to the SA pattern. The element pattern serves as the EDBF SA pattern enabling simultaneous beam scanning across the entire FOV.

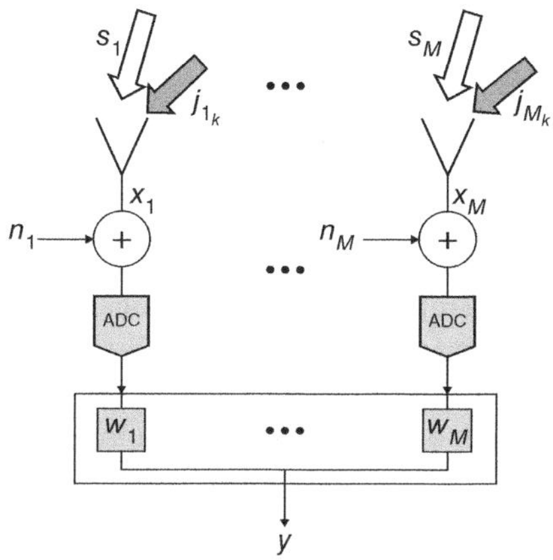

Figure 7.21 The signal x_m at the ADC is adaptively weighted to maximize signal versus noise and interference.

In Figure 7.21 the desired signal is incident on the AESA at the angle θ_0 while the jamming signal is incident at angle θ_j. An adaptive weight w_m is applied to the received signal x_m at each element to form the beam output that is the sum of all the weighted elements. x_m is inclusive of the desired signal, jamming signal(s), and environmental noise at each element. Using vector notation, this sum can be written as

$$y = \vec{w}^H \vec{x}, \tag{7.15}$$

where the superscript H is the hermitian transpose of vector $\vec{w}$. The objective of ABF is to find the optimum weights $\vec{w}$ that maximize the desired signal while minimizing interference from other directions.

The phase relationship between each element is simply the exponential term in the AF expression, which is $e^{j\frac{2\pi}{\lambda}d_m sin\theta}$. This is called the steering vector $\vec{v}(\theta)$ and can be put into vector form as

$$\vec{v}(\theta) = \left[1 \; e^{j\frac{2\pi}{\lambda}dsin\theta} \; \cdots \; e^{j\frac{2\pi}{\lambda}(M-1)dsin\theta}\right]^T, \tag{7.16}$$

where the steering vector weights phase reference is at the end of the array, $d_m = (m-1)d$, and d is the element spacing (Melvin and Scheer, 2013).

To find the optimum weights, the weight vector is optimized with respect to the power of various signals after beamforming. The most commonly used approach is to minimize the mean square error (MSE) signal or maximize the signal power relative to the jammer and noise power (Melvin and Scheer, 2013). The output power of the beamformer in Figure 7.21 can be written as (Melvin and Scheer, 2013)

$$P_{out} = E[|y|^2] = \vec{w}^H R \vec{w}, \tag{7.17}$$

where the covariance matrix R is defined as

$$\vec{R} = E[\vec{x}\vec{x}^H] = |S|^2\vec{v}(\theta_s)\vec{v}^H(\theta_s) + \sum_{k=1}^{K} |a_k|^2\vec{v}(\theta_k)\vec{v}^H(\theta_k) + \sigma_n^2\vec{I}. \quad (7.18)$$

In Equation 7.18, S is the desired signal, σ_n is the noise power, a_k is the kth jammer amplitude, and $\vec{I}$ is the identity matrix. The diagonal terms of $\vec{R}$ are the autocovariance of each element channel, and the off-diagonal terms provide information on the AOA of the signal and jammers and are adaptively exploited to minimize the jammer power incident from different angles. Minimizing the MSE of the output beamformer and the desired signal, the optimal adaptive weights can be expressed as

$$\vec{w}_{optimum} = k\vec{R}^{-1}\vec{v}(\theta_o), \quad (7.19)$$

where k is an arbitrary constant for completeness that can be set to 1 since it scales both the signal and noise (and interference) power.

Figure 7.22 illustrates a 1D 30-element AESA pattern with and without ABF. In this example, one jammer signal is incident at -30° while the main beam is scanned to 20°. A null is formed in the ABF pattern at -30°. Figure 7.23 shows the same 1D AESA with four jammers instead of one, showing the effectiveness and scalability of ABF. The adaptive weighting methodology optimally allocates its degrees of freedom (digitized channels) to the enhancement of the desired target

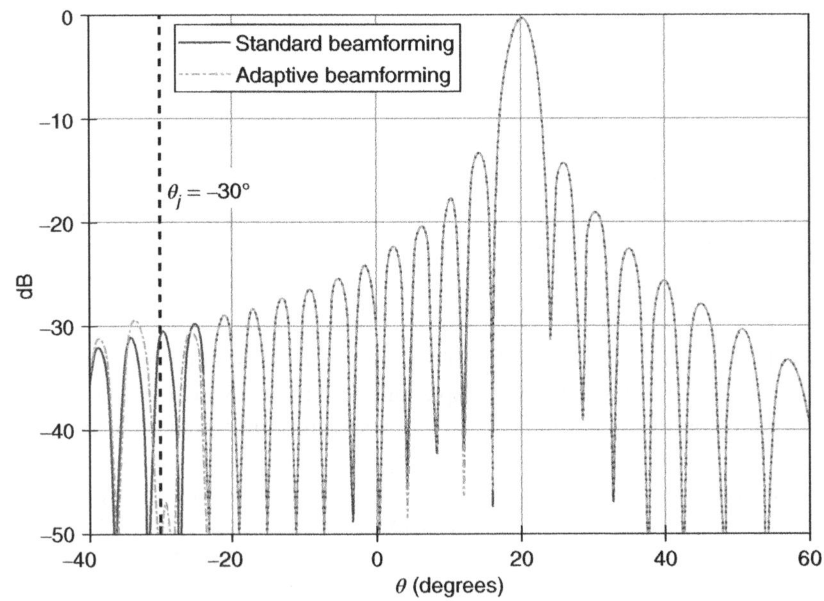

Figure 7.22 Adaptive beamformer response nulls out jammer at 30°.

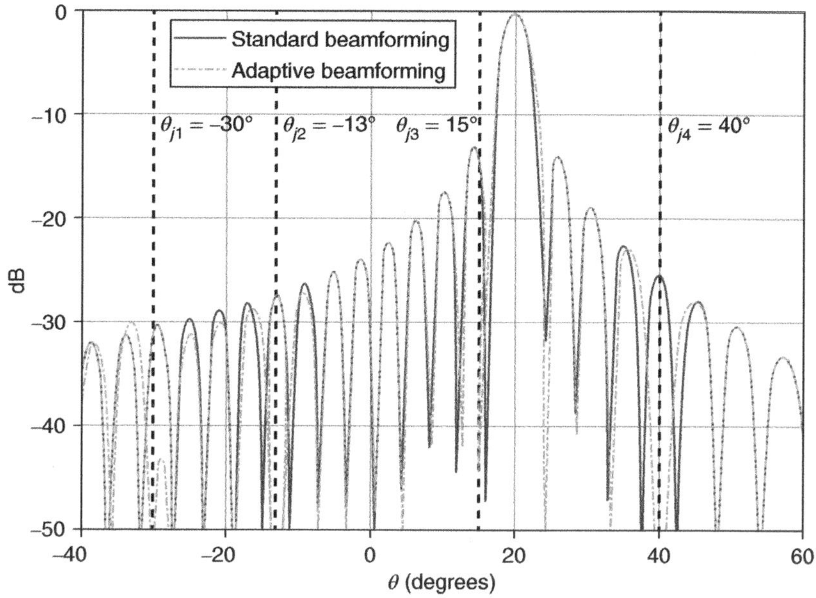

Figure 7.23 Adaptive beamformer response nulls out four jammers, demonstrating robust capability.

signal and to the cancellation of jammers and other environmental contributors such as clutter (Skolnik, 1990).

References

Melvin, W., and Scheer, J.A. *Principles of Modern Radar*. SciTech Publishing, 2013.

Skolnik, M.I. *Radar Handbook*. McGraw Hill, 1990.

Appendix A

Array Factor (AF) Derivation

In Chapter 2 the closed form for the AF was shown to be:

$$AF = \frac{sin\left[M\pi d\left(\frac{sin\theta_o}{\lambda_o} - \frac{sin\theta}{\lambda}\right)\right]}{sin\left[\pi d\left(\frac{sin\theta_o}{\lambda_o} - \frac{sin\theta}{\lambda}\right)\right]}. \tag{A.1}$$

This is derivable from the exponential summation expression for the uniform illumination AF, which is:

$$AF = \sum_{m=1}^{M} e^{j\left(\frac{2\pi}{\lambda}x_m\, sin\theta - \frac{2\pi}{\lambda_o}x_m\, sin\theta_o\right)}. \tag{A.2}$$

The position of the array elements, x_m, is expressed as $x_m = (m - \frac{M+1}{2})d_x$, where d_x is the element spacing and M is the number of array elements. Using this expression puts the phase center of the array at $x = 0$[1]. Equation (A.2) can then be rewritten as

$$\begin{aligned} AF &= \sum_{m=1}^{M} e^{jx_m\left(\frac{2\pi}{\lambda}sin\theta - \frac{2\pi}{\lambda_o}sin\theta_o\right)} \\ &= \sum_{m=1}^{M} e^{jd_x\left(m-\frac{M+1}{2}\right)\cdot\left(\frac{2\pi}{\lambda}sin\theta - \frac{2\pi}{\lambda_o}sin\theta_o\right)} \\ &= \sum_{m=1}^{M} e^{j\Psi\left(m-\frac{M+1}{2}\right)}, \end{aligned} \tag{A.3}$$

where $\psi = d_x\left(\frac{2\pi}{\lambda}sin\theta - \frac{2\pi}{\lambda_o}sin\theta_o\right)$. Equation A.3 can then be expanded as

$$\begin{aligned} AF &= \sum_{m=1}^{M} e^{j\Psi\left(m-\frac{M+1}{2}\right)} \\ &= e^{j\Psi\left(\frac{1-M}{2}\right)} + e^{j\Psi\left(\frac{3-M}{2}\right)} + \ldots + e^{j\Psi\left(\frac{M-3}{2}\right)} + e^{j\Psi\left(\frac{M-1}{2}\right)} \end{aligned} \tag{A.4}$$

1 The expression for x_m can be represented as $x_m = (m - 1)d_x$ without loss of generality; however, this gives a phase center that is offset from $x = 0$.

and when scaled by $e^{j\Psi}$ results in

$$e^{j\Psi}AF = e^{j\Psi\left(\frac{3-M}{2}\right)} + \ldots + e^{j\Psi\left(\frac{M-1}{2}\right)} + e^{j\Psi\left(\frac{M+1}{2}\right)}. \tag{A.5}$$

Subtracting Equation A.5 from Equation A.4 generates the following expression

$$AF - e^{j\Psi}AF = AF(1 - e^{j\Psi}) = e^{j\Psi\left(\frac{1-M}{2}\right)} - e^{j\Psi\left(\frac{M+1}{2}\right)}. \tag{A.6}$$

Rearranging terms in Equation A.6 results in a new expression for the AF, which is

$$AF = \frac{e^{j\Psi\left(\frac{1-M}{2}\right)} - e^{j\Psi\left(\frac{M+1}{2}\right)}}{1 - e^{j\Psi}}. \tag{A.7}$$

Equation A.7 can then be further simplified as follows

$$\begin{aligned} AF &= \frac{e^{j\Psi\left(\frac{1-M}{2}\right)} - e^{j\Psi\left(\frac{M+1}{2}\right)}}{1-e^{j\Psi}} \\ &= \frac{e^{j\left(\frac{\Psi}{2}-\frac{\Psi M}{2}\right)} + e^{\left(\frac{\Psi}{2}-\frac{\Psi M}{2}\right)}}{e^{j\frac{\Psi}{2}}\left(e^{-j\frac{\Psi}{2}} - e^{j\frac{\Psi}{2}}\right)} \\ &= \frac{e^{j\left(\frac{\Psi}{2}\right)}\left(e^{-j\left(\frac{\Psi M}{2}\right)} - e^{\left(\frac{\Psi M}{2}\right)}\right)}{e^{j\frac{\Psi}{2}}\left(e^{-j\frac{\Psi}{2}} - e^{j\frac{\Psi}{2}}\right)} \\ &= \frac{\left(e^{-j\left(\frac{\Psi M}{2}\right)} - e^{\left(\frac{\Psi M}{2}\right)}\right)}{\left(e^{-j\frac{\Psi}{2}} - e^{j\frac{\Psi}{2}}\right)}. \end{aligned} \tag{A.8}$$

Using Euler's identity $sin\theta = \frac{e^{j\theta} - e^{-j\theta}}{2j}$, Equation A.8 can then be reduced to the following expression for the AF:

$$AF = \frac{sin\left(M \cdot \frac{\Psi}{2}\right)}{sin\frac{\Psi}{2}}. \tag{A.9}$$

Substituting for Ψ we're left with:

$$AF = \frac{sin\left[M \cdot d_x\left(\frac{\pi}{\lambda}sin\theta - \frac{\pi}{\lambda_o}sin\theta_o\right)\right]}{sin\left[d_x\left(\frac{\pi}{\lambda}sin\theta - \frac{\pi}{\lambda_o}sin\theta_o\right)\right]}, \tag{A.10}$$

which is equivalent to Equation A.1.

Appendix B

Instantaneous Bandwidth Derivation

In Chapter 2, the expression for IBW was shown to be

$$IBW = \frac{kc}{Lsin\theta_o}. \quad \text{(B.1)}$$

Where k is the beamwidth factor, L is the length of the AESA, and θ_o is the maximum required scan angle. An alternative expression can be shown, which ultimately reduces to Equation B.1. We begin with the expression for the AF shown in Chapter 2:

$$AF = \sum_{m=1}^{M} a_m e^{j\left(\frac{2\pi}{\lambda}x_m sin\theta - \frac{2\pi}{\lambda_0}x_m sin\theta_o\right)}. \quad \text{(B.2)}$$

Rewriting Equation B.2 in terms of f, the AF expression becomes

$$AF = \sum_{m=1}^{M} a_m e^{j\frac{2\pi}{c}x_m(f\,sin\theta - f_o sin\theta_o)} = \sum_{m=1}^{M} a_m e^{j\frac{2\pi}{c}x_m(\Psi)}, \quad \text{(B.3)}$$

where $\Psi = (f\,sin\theta - f_o sin\theta_o)$. When operating at the tune frequency, $f = f_o$, *and* the ESA is scanned to $\theta = \theta_o$, $\Psi = 0$. However, for operation away from the tune frequency, $f = f_o + \Delta f$, $\Psi = (f_o + \Delta f)sin\theta - f_o sin\theta_o$, and Ψ is no longer zero, thus resulting in beam squint (i.e., the AF and thus the AESA pattern has a maximum value at an angle different than θ_o).

The beam squint, $\Delta\theta$, can be calculated by substituting $f_o + \Delta f$ and $\theta = \theta_o + \Delta\theta$ into the expression for Ψ:

$$\begin{aligned} \Psi &= f\,sin\theta - f_o \sin\theta_o \\ &= (f_o + \Delta f)sin(\theta_o - \Delta\theta) - f_o sin\theta_o. \end{aligned} \quad \text{(B.4)}$$

It is important to note that for $\Delta f > 0$, $\Delta\theta < 0$, meaning the beam squints to a value less than the scan angle for off-tune frequencies greater than the tune frequency. For off-tune frequencies less than the tune frequency, $\Delta f < 0$, $\Delta\theta > 0$. Setting Ψ equal to 0 in Equation B.4, applying a trigonometric identity

Active Electronically Scanned Arrays: Fundamentals and Applications, First Edition. Arik D. Brown.

($sin(A - B) = sinAcosB - sinBcosA$), and solving for $\Delta\theta$ produces the following expression:

$$(f_o + \Delta f)\left[sin(\theta_o)cos(\Delta\theta) - sin(\Delta\theta)cos(\theta_o)\right] = f_o sin\theta_o. \quad \text{(B.5)}$$

Using small-angle approximations for the beam squint terms ($sin\alpha \approx \alpha$ and $cos\alpha \approx 1$), reduces Equation B.5 to

$$\Delta\theta = \frac{\Delta f}{(f_o + \Delta f)} tan\theta_o \quad \text{(B.6)}$$

$$\approx \frac{\Delta f}{f_o} tan\theta_o. \quad \text{(B.7)}$$

The expression in B.7 is the same as what is found in (Skolnik, 1990). Recognizing that Δf is the IBW, and substituting the expression for the beamwidth for the beam squint $\left(\Delta\theta = kBW = k\left(\frac{\lambda}{L}\right)\right)$, we arrive at the following equation:

$$\begin{aligned} IBW &= \frac{k\lambda}{L} \cdot \frac{f_o}{tan\theta_o} \\ &= \frac{kc}{Ltan\theta_o}. \end{aligned} \quad \text{(B.8)}$$

For scan angles less than 20°, Equation B.8 is identical in value to Equation B.1.

Reference

Skolnik, M.I. *Radar Handbook*. McGraw Hill, 1990.

Appendix C

Triangular Grid Grating Lobes Derivation

In Chapter 2, the grating lobe expression for a triangular grid of elements was shown to be (Skolnik, 1990), (Corey, 1985):

$$u_m = u_o + m\frac{\lambda}{2d_x},\ v_n = v_o + n\frac{\lambda}{2d_y}$$
$$m, n = 0, \pm1, \pm2, \cdots \quad m + n \ is \ even. \tag{C.1}$$

This expression is derived in (Corey, 1985); however, a more intuitive derivation will be shown here. We begin by considering a rectangular array with element spacing $2d_x$ and $2d_y$, as shown in Figure C.1. In Chapter 2 it was shown that the grating lobes for an array with this element spacing will occur at

$$\begin{aligned} u_m &= u_o + m\frac{\lambda}{2d_x},\ m = 0, \pm1, \pm2, \cdots \\ v_n &= v_o + n\frac{\lambda}{2d_y},\ n = 0, \pm1, \pm2, \cdots \end{aligned} \tag{C.2}$$

Next, consider another array of elements with the same $2d_x$ and $2d_y$ spacing, but offset by d_x and d_y, as shown in Figure C.2. This offset can be expressed mathematically with a complex exponential phase shift: $e^{-j\frac{2\pi}{\lambda}[d_x(u-u_o)+d_y(v-v_o)]}$. The combined array factor (AF_{total}) for the two offset rectangular arrays can be expressed as

$$\begin{aligned} AF_{total} &= AF_1 + AF_2 \\ &= AF_1 + AF_1 e^{-j\frac{2\pi}{\lambda}[d_x(u-u_o)+d_y(v-v_o)]} \\ &= (1 + e^{-j\frac{2\pi}{\lambda}[d_x(u-u_o)+d_y(v-v_o)]})AF_1. \end{aligned} \tag{C.3}$$

In Equation C.3, AF_1 and AF_2 are the array factors for the two arrays, respectively. From Equation C.2, the AF in Equation C.1 has maxima (i.e., grating lobes) that

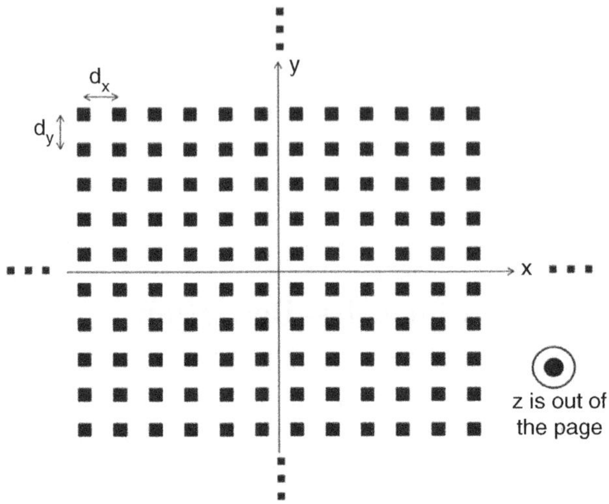

Figure C.1 Rectangular grid element spacing establishes the location of grating lobes in sine space.

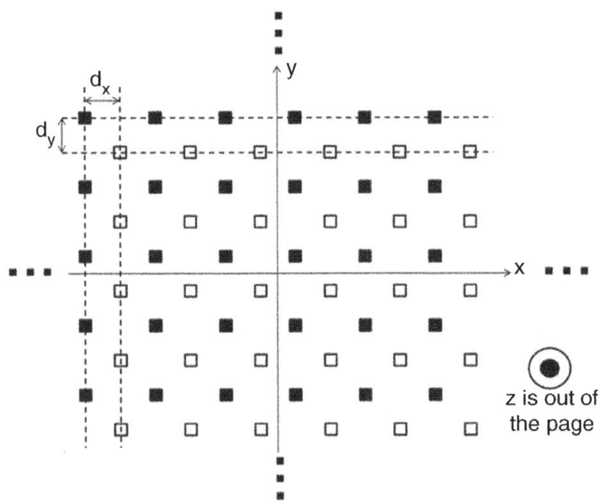

Figure C.2 An AESA with a triangular grid can be represented as two rectangular grids offset by d_x and d_y.

occur at multiples of $m\frac{\lambda}{2d_x}$ and $n\frac{\lambda}{2d_y}$. Substituting into the complex exponential phase shift term in Equation C.3 gives

$$\begin{aligned} 1+e^{-j\frac{2\pi}{\lambda}\left(d_x\frac{m\lambda}{2d_x}+d_y\frac{n\lambda}{2d_y}\right)} &= 1+e^{-j\pi[m+n]} \\ &= \begin{cases} 0, \mathit{for}\ m+n\ \mathit{odd} \\ 1, \mathit{for}\ m+n\ \mathit{even} \end{cases}. \end{aligned} \tag{C.4}$$

From Equation C.4 we see that the AF expression in Equation C.3 has maxima only when the quantity $m+n$ is even. These maxima are the grating lobes for a triangular grid, which is a superposition of two rectangular grids spatially offset. The expression for the grating lobe locations in sine space is then

$$\begin{aligned} u_m &= u_o + m\frac{\lambda}{2d_x}, \quad m = 0, \pm1, \pm2, \cdots \\ v_n &= v_o + n\frac{\lambda}{2d_y}, \quad n = 0, \pm1, \pm2, \cdots \\ &\mathit{for}\ m+n\ \mathit{even}. \end{aligned} \tag{C.5}$$

Equation C.5 is equivalent to Equation C.1.

References

Skolnik, M.I. *Radar Handbook*. McGraw Hill, 1990.

Corey, L. "A method for minimizing the number of elements in a phased-array antenna." *Antennas and Propagation Society International Symposium*, pp. 241–244, 1985.

Appendix D

General Expression for Intercept Point Derivation

The general expression for the intercept point, as discussed in Chapter 4, can be derived in a few simple steps. This expression can be used to calculate the intercept point for nth-order spurious products generated by a nonlinear device such as an amplifier. It is important to point out that the equations shown are decibel values. Two scalars multiplied together results in their decibel values being *added* together, i.e., ab will be $10log_{10}(a) + 10log_{10}(b) = A + B$. This is a subtle point but important for understanding the following derivation.

Equation 2.6 expresses the output nth-order power in decibels as

$$P_{o_n} = nP_{in} + c_1, \tag{D.1}$$

where P_{o_n} is the nth-order output power, n is the order of the nonlinear output, P_{in} is the linear input power, and c_1 is an arbitrary constant. P_{o_1} can be expressed as

$$P_{o_1} = P_{in} + G, \tag{D.2}$$

where G is the gain of the nonlinear device. Substituting Equation D.2 in Equation D.1 results in a modified expression for P_{o_n}, which is

$$\begin{aligned} P_{o_n} &= n(P_{o_1} - G) + c_1 \\ &= nP_{o_1} - nG + c_1 \\ &= nP_{o_1} + c_2, \quad c_2 = -nG + c_1. \end{aligned} \tag{D.3}$$

By definition, at the nth-order intercept point, IPn, $P_{o_n} = P_{o_1} = IP_n$, and Equation D.3 becomes:

$$IP_n = nIP_n + c_2. \tag{D.4}$$

Rearranging Equation D.4, c_2 can be expressed as

$$c_2 = IP_n(1 - n). \tag{D.5}$$

Substituting c_2 into Equation D.3 gives

$$P_{o_n} = nP_{o_1} + (1 - n)IP_n, \tag{D.6}$$

Active Electronically Scanned Arrays: Fundamentals and Applications, First Edition. Arik D. Brown.

which is the expression for the nth-order output power as a function of the output linear power and intercept point. Equation D.6 can be rearranged to show the expression for the IP_n, which is

$$IP_n = \frac{1}{(1-n)} P_{o_n} - \frac{n}{(1-n)} P_{o_1}. \tag{D.7}$$

Using Equation D.6, the third-order intercept point will be calculated as an example. Setting $n = 3$, Equation D.6 becomes

$$\begin{aligned} P_{o_3} &= 3P_{o_1} + (1-3)IP_3 \\ &= 3P_{o_1} + (-2)IP_3 \\ &= 3\left(P_{o_1} - \tfrac{2}{3}IP_3\right), \end{aligned} \tag{D.8}$$

which is the familiar expression for P_{o_3}. Equation D.8 is used to calculate the spurious free dynamic range (SFDR) relative to the IP_3 (see Chapter 6). The maximum P_{o_1} is determined when P_{o_3} is equal to the noise floor, $kTBGF$. In practice, P_{o_3} is set to a lower value to ensure spurious products are kept below the noise floor.

Appendix E

Impact of Failed Elements on AESA Performance

In Chapter 4, the impact of failed elements on AESA reliability was discussed. In addition to reliability, failed TRMs also affect the signal power in the RRE. Failed TRMs reduce P_{TX} and G_{TX}, which are the primary drivers of radiated power and transmit receive (TR) gain. For different missions, the effects on the performance are represented in Table E.1. This appendix will quantitatively capture these effects using the RRE. It is assumed for the following derivations that each TRM represents a single channel/element allowing TRM and element to be used interchangeably.

Consider an AESA that has F failed elements out of N total TRMs. The number of operational TRMs can be represented as:

$$operational\ TRMs = N - F. \tag{E.1}$$

Using Equation E.1, P_{TX}, can be represented as:

$$P_{TX} = P_E(N - F), \tag{E.2}$$

where P_E is the transmitted power per TRM. Similarly, G_{TX} can modified to include the effects of failed TRMs with the following expression:

$$G_{TX} = \frac{4\pi A}{\lambda^2} = \frac{4\pi A_E}{\lambda^2}(N - F), \tag{E.3}$$

where A_E is the effective area per element. Using Equations E.2 and E.3, the ERP for the AESA can be represented as:

$$ERP = P_{TX}G_{TX} = P_E\frac{4\pi A_E}{\lambda^2}(N - F)^2. \tag{E.4}$$

Equation E.4 is very useful as it quantifies the effect of failed TRMs on the AESA's ERP and shows that it is quadratic: $(N - F)^2$. As will be shown later, Equation E.4 can be manipulated and used to express the loss in ERP in terms of the percentage of failed elements, $\left(100 \cdot \frac{F}{N}\right)$. This will directly quantify the performance effect for applications employing AESA's for EA.

Active Electronically Scanned Arrays: Fundamentals and Applications, First Edition. Arik D. Brown.

Table E.1 AESA failed elements have an effect on system performance for various missions.

Mission	System Performance Impact
Radar	maximum detection range
EA	maximum ERP for jammer-to-signal (J/S) ratio
ESM	receive sensitivity
Comms	link closure, bit error rate (BER)

Next, the received signal power in the RRE will be expressed in terms of the number of failed elements. Using Equation E.4, the reflected signal power in the RRE can be expressed as

$$\begin{aligned} S &= \frac{ERP}{4\pi R^2} \cdot \sigma \cdot \frac{1}{4\pi R^2} \frac{\lambda^2 G_{RX}}{4\pi} \\ &= ERP \cdot G_{RX} \cdot \frac{\sigma \lambda^2}{(4\pi)^3 R^4} \\ &= P_E \left(\frac{4\pi A_E}{\lambda^2} \right)^2 (N-F)^3 \cdot \frac{\sigma \lambda^2}{(4\pi)^3 R^4}, \end{aligned} \tag{E.5}$$

where $G_{RX} = G_{TX}$. Equation E.5 shows that there is a cubic relation between degraded signal performance and failed elements in the form of $(N-F)^3$.

Thus far the AESA signal power degradation has been shown for transmit (Equation E.4) and TR (Equation E.5) applications. Next, the effect on receive-only applications will be shown. This is done by starting with the received signal power for receive-only operation and then substituting for G_{RX} ($G_{RX} = G_{TX}$) using Equation E.3.

$$\begin{aligned} S &= \frac{ERP_{external}}{4\pi R^2} \frac{\lambda^2 G_{RX}}{4\pi} \\ &= ERP_{external} \cdot \left(\frac{\lambda}{4\pi R} \right)^2 \cdot G_{RX} \\ &= ERP_{external} \cdot \left(\frac{\lambda}{4\pi R} \right)^2 \cdot \frac{4\pi A_E}{\lambda^2} (N-F). \end{aligned} \tag{E.6}$$

In Equation E.6, $ERP_{external}$ is used to denote that this ERP is not from the AESA but is from a signal in the environment. Thus, it is not modified with Equation E.4. Equation E.6 shows that for receive-only applications, the degradation to signal power is linear: $(N-F)$.

Using Equations E.4, E.5, and E.6, the effect of failed elements on AESA signal power can be summarized. Before doing this, it is useful to modify the expression

Table E.2 AESA signal power loss expressed as a function of the number of failed elements, *F*.

Application	AESA Signal Power Loss	*k*	Mission
Receive only	$\left(1-\frac{F}{N}\right)^k$	1	ESM, comms
Transmit only	$\left(1-\frac{F}{N}\right)^k$	2	EA, comms
Transmit and receive	$\left(1-\frac{F}{N}\right)^k$	3	radar

for operational TRMs to provide additional insight.

$$\begin{aligned} Operational\ TRMs &= N - F \\ &= N\left(1-\frac{F}{N}\right). \end{aligned} \tag{E.7}$$

Equation E.7 is useful because the term $\left(1-\frac{F}{N}\right)$ can be used to express the AESA signal power loss as

$$AESA\ signal\ power\ loss = \left(1-\frac{F}{N}\right)^k,\ k = 1, 2, or\ 3 \tag{E.8}$$

and is summarized in Table E.2 and illustrated in Figure E.1.

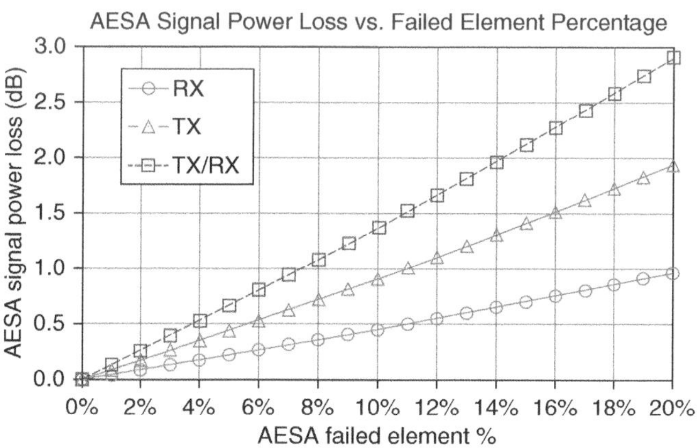

Figure E.1 AESA signal power loss expressed as a function of percentage of failed elements.

Appendix F

Sidelobe Blanking with an AESA

In Chapter 5, it was shown that a beamformer using 180° hybrid couplers could create Σ, Δ_{AZ}, and Δ_{EL} array beams for performing monopulse AoA. The fourth port, although typically loaded, produces what is referred to as a $\Delta\Delta$ beam. This type of beam, for much of the AESA field of view, has a pattern that in a majority of the spatial FOV is higher than that of the Σ beam sidelobes (see Figures 5.23 and 5.24). It can be used as an auxiliary output that can help reject intentional (jammers) or unintentional interference.

To implement sidelobe blanking, an additional receiver channel is required. Thus, for a 2D monopulse AESA with an auxiliary output, four channels are necessary. The main beam (main beam is equivalent to the Σ beam) and auxiliary beam are sent to the receiver for comparison (Skolnik, 1990). In practice, a single array element from the AESA can be used for the auxiliary channel. The array element has a broad omnidirectional pattern in comparison to the main beam providing coverage over the SL region. This is illustrated in Figure F.1.

A 900-element AESA is assumed that is 30 × 30 array elements in size. A 30 dB, nbar = 5, Taylor weight is applied to reduce the sidelobes. The taper loss for the 2D distribution is ~ 1.36 dB. A single element in the AESA is used for the auxiliary channel. Figure F.1 shows an azimuth pattern cut ($\phi = 0°$ and $\theta = -90°$ to $+90°$) of both the main and auxiliary beams. The patterns in Figure F.1 are normalized to the peak of the main beam. This is done because the sidelobe blanking comparison requires the comparison of the array gains. For analysis, the array gain of the main beam can be expressed as:

$$\begin{aligned} G_{a_{main}} &= \frac{4\pi A_{main}}{\lambda^2} \cdot TL \\ &= \frac{4\pi M A_{auxiliary}}{\lambda^2} \cdot TL, \end{aligned} \tag{F.1}$$

where A_{main} is the effective area of the AESA, M is the number of array elements, $A_{auxiliary}$ is the effective area of a single array element, and TL is the taper loss as defined in Chapters 2 and 5. A similar expression can be written for the array gain

Active Electronically Scanned Arrays: Fundamentals and Applications, First Edition. Arik D. Brown.

of a single element as

$$G_{a_{auxiliary}} = \frac{4\pi A_{auxiliary}}{\lambda^2}. \tag{F.2}$$

The ratio of the gains in Equations F.1 and F.2 is the ratio of power between the beams and can be expressed as

$$\frac{G_{a_{main}}}{G_{a_{auxiliary}}} = \frac{\frac{4\pi M A_e}{\lambda^2} \cdot TL}{\frac{4\pi A_{auxiliary}}{\lambda^2}} = M \cdot TL. \tag{F.3}$$

Expressing Equation F.3 in dB gives

$$\begin{aligned} \left[\frac{G_{a_{main}}}{G_{a_{auxiliary}}}\right]_{dB} &= 10log_{10}(M) - 10log_{10}(TL) \\ &= M_{dB} - TL_{dB} \end{aligned} \tag{F.4}$$

For the patterns shown in Figure F.1 the dB difference is $\simeq$ 28 dB ($M_{dB} - TL_{dB} = 10log_{10}(900) - 10log_{10}(1.36) \simeq 28$ dB).

Figure F.2 shows the main beam scanned to 50°. Since the auxiliary element does not scan, its pattern remains fixed. This provides an opportunity for some interference to enter the SLs that have a power level above the auxiliary pattern. This can be mitigated by using a larger taper at the cost of array gain sensitivity.

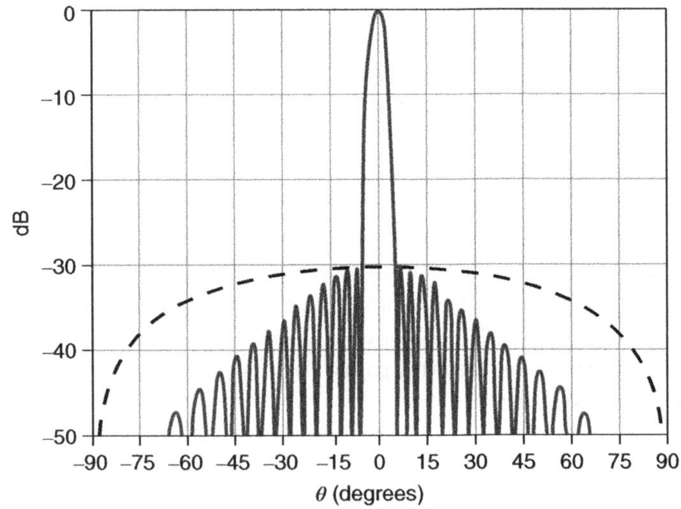

Figure F.1 Single element from the AESA can be used for the auxiliary channel for SL blanking. The pattern of a single element is omni-directional covering the SLs of the main beam. The patterns in the figure show an azimuth cut, $\phi = 0^o$ and $\theta = -90^o$ to $+90^o$, for both the main array and auxiliary element.

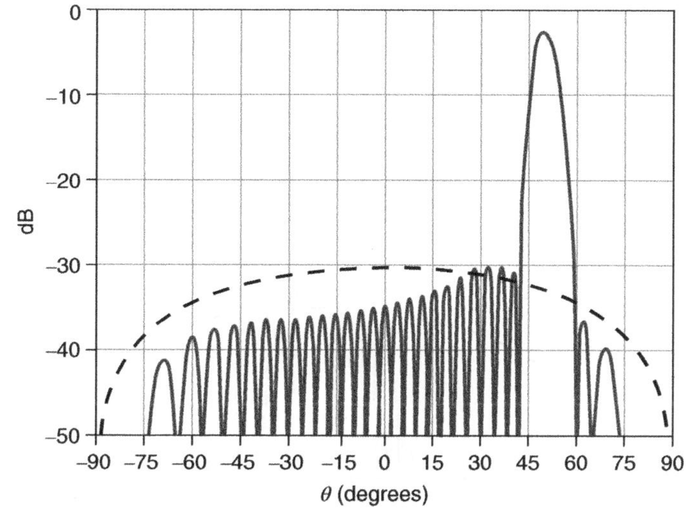

Figure F.2 When the main beam is scanned, some of the close-in SLs are higher than the auxiliary element pattern. The single element used for the auxiliary channel does not scan and has some degradation in SL blanking performance as the scan angle is increased.

The patterns shown in Figures F.1 and F.2 do not include errors for the main beam. In an AESA, as shown in Chapter 2, there will be average amplitude and phase errors. This causes the SLs to have irregular behavior. Figures F.3 and F.4 show the main and auxiliary beams for the current AESA example with amplitude (± 2 dB) and phase ($\pm 3^o$) errors included for the main beam. (Since the auxiliary element is one element, amplitude and phase errors do not need to be included for its array pattern.) For the boresite case with errors, there is some degradation near the main beam. However, for the scanned case with errors, there is an increased number of SLs that exceed the auxiliary pattern by a larger amount than the boresite case. The SLs that scan toward 0° become larger since they are no longer reduced as much by the element pattern near boresite. The SLs near the main beam decrease in amplitude, but so does the element pattern. These two conditions provide a challenge for sidelobe blanking with an AESA. For an MSA, performance always mimics that of the boresite case and hence does not have this feature. To overcome this, a combination of minimizing amplitude and phase errors for the AESA, increasing the SL taper, and/or non-adaptive or adaptive nulling provide a complement of solutions for mitigation.

The current example thus far assumes that the noise power for both the main and auxiliary beams are equal, and the only difference is array gain. This, though, is typically not the case. As previously shown with the ratio of main beam gain

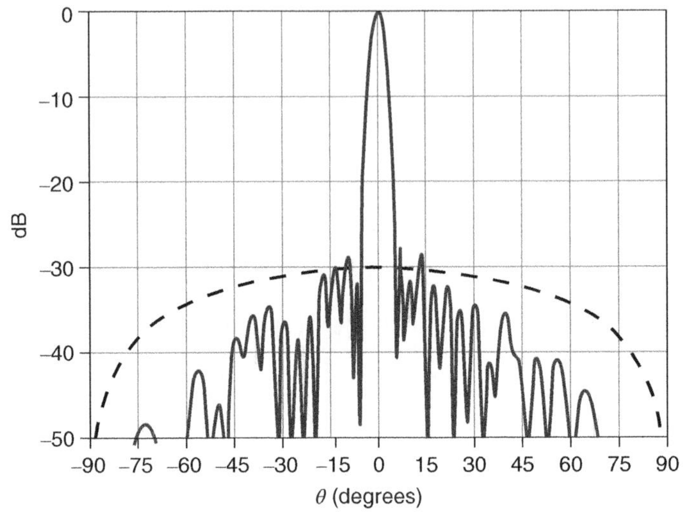

Figure F.3 Amplitude and phase errors create higher SLLs for AESAs. This cause SLs to rise above the auxiliary element pattern even at boresite.

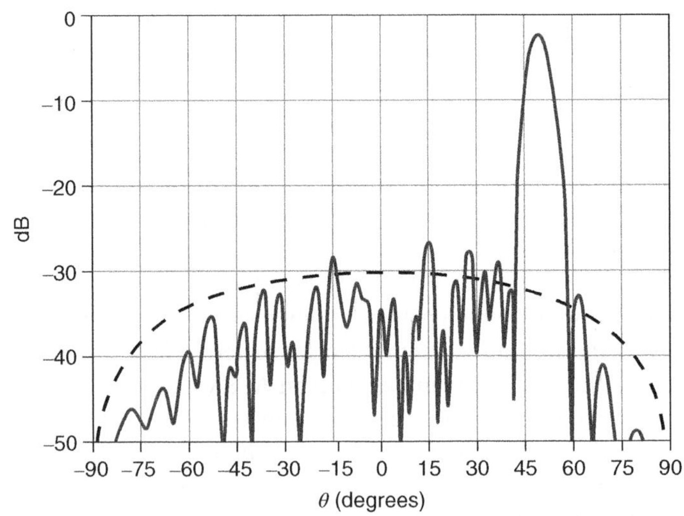

Figure F.4 Similar to Figure F.3, amplitude and phase errors cause higher SLLs for AESAs and can further degrade performance for SL blanking when the main beam is scanned.

to auxiliary gain, a similar expression between the main and auxiliary SNRs will provide a slightly modified result. To rigorously calculate the difference between the signal and noise power for the main and auxiliary beams, the signal and noise voltages are used. However, the resulting expression simplifies to a comparison of the SNRs between the two beams.

To begin, the SNR for the main beam is expressed as

$$\begin{aligned} SNR_{main} &= \frac{ERP_{external}\lambda^2 G_{a_{main}} G_{elec_{main}}}{(4\pi R)^2} \cdot \frac{1}{kT_{main} G_{elec_{main}} BL} \\ &= \frac{ERP_{external}\lambda^2 G_{a_{main}}}{(4\pi R)^2} \cdot \frac{1}{kT_o F_{main} BL}, \end{aligned} \tag{F.5}$$

where $ERP_{external}$ is $\frac{P_{a_{main}} G_{a_{main}} \sigma}{4\pi R^2}$ for radar and $P_{external} G_{external}$ for receive only, and G_{elec} is the electronic gain. Similarly, the SNR for the auxiliary channel is

$$SNR_{auxiliary} = \frac{ERP_{external}\lambda^2 G_{a_{auxiliary}}}{(4\pi R)^2} \cdot \frac{1}{kT_o F_{auxiliary} BL}. \tag{F.6}$$

Expressing the SNRs as a ratio results in the following equation

$$\begin{aligned} \frac{SNR_{main}}{SNR_{auxiliary}} &= \frac{\frac{ERP_{external}\lambda^2 G_{a_{main}}}{(4\pi R)^2} \cdot \frac{1}{kT_o F_{main} BL}}{\frac{ERP_{external}\lambda^2 G_{a_{auxiliary}}}{(4\pi R)^2} \cdot \frac{1}{kT_o F_{auxiliary} BL}} \\ &= \frac{G_{a_{main}}}{G_{a_{auxiliary}}} \cdot \frac{F_{auxiliary}}{F_{main}}. \end{aligned} \tag{F.7}$$

There are two conditions of interest for which Equation F.7 sheds insight. First, in the main beam region, as previously shown, the main beam gain is significantly larger than than the auxiliary beam gain. This means that the ratio of SNRs will always be large and positive in the 3 dB region. Second, in the SL region, $G_{a_{main}}$ and $G_{a_{auxiliary}}$ are similar in amplitude. However, since the auxiliary element is a single element, its noise figure ($NF_{auxiliary}$) will be less than that of the main beam, making the SNR ratio small. Mathematically, this can be depicted as

$$\begin{aligned} \frac{SNR_{main}}{SNR_{auxiliary}} &= \frac{G_{a_{main}}}{G_{a_{auxiliary}}} \cdot \frac{F_{auxiliary}}{F_{main}} \\ &= \frac{G_{a_{main}}}{G_{a_{auxiliary}} \cdot \frac{F_{main}}{F_{auxiliary}}}. \end{aligned} \tag{F.8}$$

By setting an appropriate threshold between the main and auxiliary channels as described in (Skolnik, 1990), main beam SLs can be blanked out even if they are slightly larger in power than the auxiliary pattern. This reduces some of the burden previously described on minimizing amplitude and phase errors in the AESA, increasing the taper, or adding complexity with adaptive nulling.

Reference

Skolnik, M.I. *Radar Handbook*. McGraw Hill, 1990.

Appendix G

External Noise Considerations[1]

In Section 6.3.2, the AESA output noise power was formulated. The output noise is a function of the external noise temperature $T_{external}$ presented to the AESA by the environment. In Chapter 6, $T_{external}$ was expressed as T. In some situations, a more detailed representation of T as a function of $T_{external}$ is required. As an example, a ground-based AESA pointing toward the sky looking for targets in space will be considered.

Assume an AESA system looking through clouds for a target in space. Three cloud layers will be assumed represented by the letters A, B, and C. The clouds have a loss of L_A, L_B, and L_C, respectively, where each loss term L is greater than or equal to one. As an example, for a loss of $\frac{1}{2}$, $L = 2$. Having defined the cloud losses, the noise presented to the LNA in the TRM can be represented by the illustration in Figure G.1. $T_{external}$ is the noise temperature above the three cloud layers and T_A, T_B, and T_C are the noise temperatures of the individual cloud layers. The last block denoted by the subscript E represents the pre-LNA noise temperature and losses in the array element in addition to any losses that occur post-array element and pre-LNA. Also, the noise bandwidth is assumed to have a value of 1 for ease of illustration (i.e., $B = 1$ Hz).

In Chapter 6, it was shown that the output noise of an attenuator is equal to the input noise. This is shown in Figure G.2. Figure G.2 has the underlying assumption that the input noise of the attenuator and its internal noise are the same. (This is a valid assumption for the attenuators internal to an AESA.) If these temperatures are not the same, then the noise model in Figure G.2 must be modified, as shown in Figure G.3. Using this modified attenuator noise model, the noise temperature presented to the AESA can be derived.

1 This appendix is based on a technical note from Mr. Bill Hopwood.

Active Electronically Scanned Arrays: Fundamentals and Applications, First Edition. Arik D. Brown.

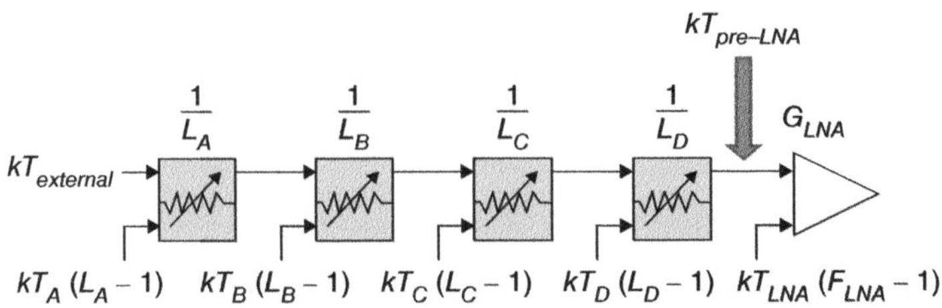

Figure G.1 Noise temperature model used to calculate the effective noise temperature presented to the LNA ($kT_{pre-LNA}$). This example assumes three cloud layers in the sky that are between an AESA system and space.

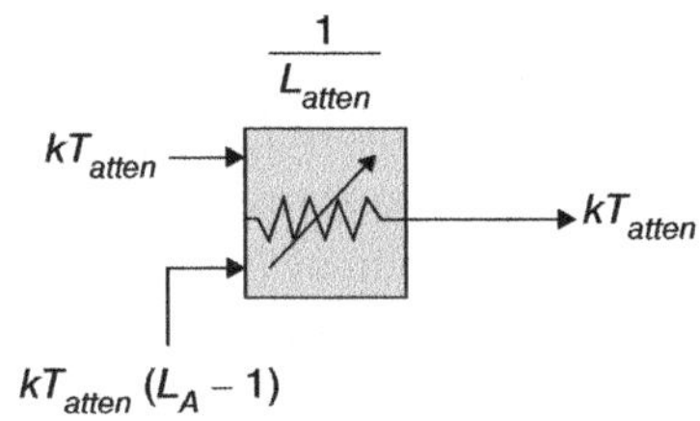

Figure G.2 Noise model for an attenuator. When the external temperature and the attenuator temperature are equal, the noise output power equals the noise input power.

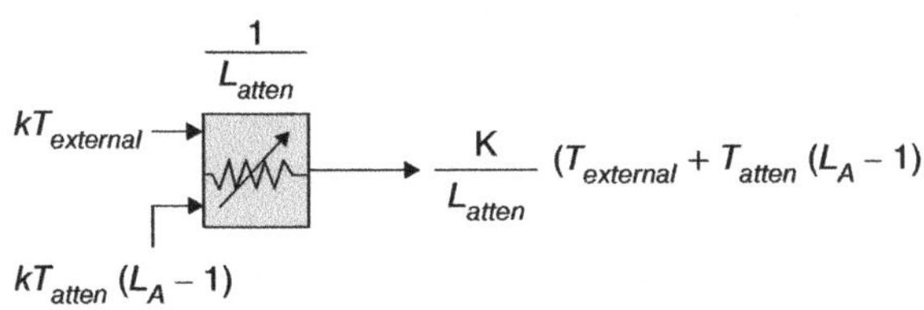

Figure G.3 Noise model for an attenuator when the external temperature is not the same as the attenuator temperature. When $T_{external} = T_{atten}$, the output noise power reduces to kT_{atten}, which is consistent with Figure G.2.

In Figure G.1, the noise presented to the LNA can be expressed as

$$N_{pre-LNA} = \frac{kT_{external}}{L_A L_B L_C L_D} + \frac{kT_A(L_A-1)}{L_A L_B L_C L_D} + \frac{kT_B(L_B-1)}{L_B L_C L_D} + \frac{kT_C(L_C-1)}{L_C L_D} + \frac{kT_D(L_D-1)}{L_D}. \tag{G.1}$$

From Equation G.1, several observations can be made. First, if all of the losses are equal to 1, implying no loss prior to the LNA, then Equation G.1 reduces to

$$N_{pre-LNA} = kT_{external}, \tag{G.2}$$

which shows the pre-LNA noise reduces to $kT_{external}$. Second, if all the losses equal 1 except for the array element loss ($L_A = L_B = L_C = 1$), then Equation G.1 can be expressed as

$$\begin{aligned} N_{pre-LNA} &= \frac{kT_{external}}{L_D} + \frac{kT_D(L_D - 1)}{L_D} \\ &= \frac{k}{L_D}\left(T_{external} + T_D(L_D - 1)\right). \end{aligned} \tag{G.3}$$

This is the same result for the noise model depicted in Figure G.3. If $T_{external} = T_D$, then $N_{pre-LNA} = kT_{external}$. This means that if the atmospheric losses are small, $L \approx 1$, and the array elements are at the same temperature as $T_{external}$, then the noise temperature seen by the LNA is $T_{external}$. Finally, by rearranging terms in Equation G.1, an effective noise temperature can be formulated. This is shown in Equation G.4

$$\begin{aligned} N_{pre-LNA} &= \frac{kT_{external}}{L_A L_B L_C L_D} + \frac{kT_A(L_A - 1)}{L_A L_B L_C L_D} + \frac{kT_B(L_B - 1)}{L_B L_C L_D} \\ &\quad + \frac{kT_C(L_C - 1)}{L_C L_D} + \frac{kT_D(L_D - 1)}{L_D} \\ &= \frac{k}{L_A L_B L_C L_D}\left(T_{external} + T_A(L_A - 1) + L_A T_B(L_B - 1)\right. \\ &\quad \left. + L_A L_B T_C(L_C - 1) + L_A L_B L_C T_D(L_D - 1)\right) \\ &= \frac{k}{L_A L_B L_C L_D} T_{pre-LNA}. \end{aligned} \tag{G.4}$$

Appendix H

Important AESA Equations Reference

H.1 System Level Equations

H.1.1 Radar Range Equation

Transmit and Receive (Radar)

$$\begin{aligned} SNR_{2\ way} &= \frac{P_{TX}G_{TX}A_{RX}\sigma\lambda^2}{(4\pi)^2R^4kTBL} \\ &= \frac{P_{TX}G_{TX}^2\sigma\lambda^2}{(4\pi)^3R^4kTBL}. \end{aligned} \tag{H.1}$$

Receive Only (ESM/SIGINT, Comms)

$$\begin{aligned} SNR_{1\ way} &= \frac{ERP_{external}A_{RX}}{4\pi R^2kTBL} \\ &= ERP_{external}G_{RX}\left(\frac{\lambda}{4\pi R}\right)^2\frac{1}{kTBL} \end{aligned} \tag{H.2}$$

$$ERP_{external} = P_{external}G_{external}. \tag{H.3}$$

Transmit Only (EA, Comms)

$$Signal\ Density = \frac{P_{TX}G_{TX}}{4\pi R^2} = \frac{ERP_{TX}}{4\pi R^2}. \tag{H.4}$$

H.1.2 Signal and Noise Gain

$$S = G_{array}G_{electronic}, \tag{H.5}$$

$$N = kTBG_{electronic}, \tag{H.6}$$

$$T = T_{external} + T_o(F-1), \tag{H.7}$$

$$T\big|_{T_{external}=T_o} = T_oF. \tag{H.8}$$

Active Electronically Scanned Arrays: Fundamentals and Applications, First Edition. Arik D. Brown.

H.1.3 Array Gain

$$D = \frac{4\pi A}{\lambda^2}, \tag{H.9}$$

$$G_{array} = \epsilon D, \quad \epsilon \leq 1. \tag{H.10}$$

H.2 AESA Theory

H.2.1 1D Pattern

Definitions:

- F = Pattern
- EP = Element Pattern
- AF = Array Factor

H.2.1.1 Phase Shifter and Time Delay Steering

Element pattern is assumed to be identical for all M elements. a_m is an amplitude weight for SL control.

Pattern Multiplication

$$F = EP \cdot AF. \tag{H.11}$$

Phase Shifter Steering

$$F(\theta) = \cos^{\frac{EF}{2}}\theta \cdot \sum_{m=1}^{M} a_m e^{j\left(\frac{2\pi}{\lambda} x_m \sin\theta - \frac{2\pi}{\lambda_o} x_m \sin\theta_o\right)}. \tag{H.12}$$

Time Delay Steering

$$F(\theta) = \cos^{\frac{EF}{2}}\theta \cdot \sum_{m=1}^{M} a_m e^{j\frac{2\pi}{\lambda} x_m (\sin\theta - \sin\theta_o)}. \tag{H.13}$$

H.2.1.2 General Expression

Element pattern is not assumed to be identical for all M elements, and pattern multiplication does not apply.

$$F(\theta) = \sum_{m=1}^{M} a_m EP_m(\theta) \cdot e^{j\left(\frac{2\pi}{\lambda} x_m \sin\theta - \frac{2\pi}{\lambda_o} x_m \sin\theta_o\right)}. \tag{H.14}$$

H.2.1.3 Conformal Array

$$F(\mathbf{r}) = \sum_{m=1}^{M} a_m (\hat{n} \cdot \hat{r})^{\frac{EF}{2}} e^{jk\mathbf{r}_i \cdot \hat{\mathbf{r}}} e^{-jk_o \mathbf{r}_i \cdot \hat{\mathbf{r}}_o}. \tag{H.15}$$

H.2.1.4 Alternate AF Expression

$$AF = \frac{sin\left[M\pi d\left(\frac{sin\theta_o}{\lambda_o} - \frac{sin\theta}{\lambda}\right)\right]}{sin\left[\pi d\left(\frac{sin\theta_o}{\lambda_o} - \frac{sin\theta}{\lambda}\right)\right]}. \tag{H.16}$$

H.2.2 2D Pattern

The element pattern is assumed the same for each element, and the AESA is planar (element locations are in the xy-plane).

Rectangular Grid

$$\begin{aligned} F(\theta,\phi) &= cos^{\frac{EF}{2}}\theta \sum_{m=1}^{M} a_m e^{j\Phi_m}, \\ \Phi_m &= e^{j\left[\left(\frac{2\pi}{\lambda}x_m\, sin\theta\, cos\phi + \frac{2\pi}{\lambda}y_m\, sin\theta\, sin\phi\right) - \left(\frac{2\pi}{\lambda_o}x_m\, sin\theta_o\, cos\phi_o + \frac{2\pi}{\lambda_o}y_m\, sin\theta_o\, sin\phi_o\right)\right]}. \end{aligned} \tag{H.17}$$

Circular Grid

$$F(\theta,\phi) = cos^{\frac{EF}{2}}\theta \cdot \sum_{k=1}^{K}\sum_{p_k=1}^{P_k} c_{k,p_k} e^{j\left[\left(\frac{2\pi}{\lambda}x_{k,p_k}u + \frac{2\pi}{\lambda}y_{k,p_k}v\right) - \left(\frac{2\pi}{\lambda_o}x_{k,p_k}u_o + \frac{2\pi}{\lambda_o}y_{k,p_k}v_o\right)\right]}. \tag{H.18}$$

H.2.3 Beamwidth

k is the beamwidth factor and for a uniform distribution ($a_m = 1$ for $m = 1, 2, \cdots, M$), $\theta_{BW_{3dB}} = 0.886$, and $\theta_{BW_{4dB}} = 1$.

$$\theta_{BW} = \frac{k\lambda}{Lcos\theta_o}. \tag{H.19}$$

H.2.4 Instantaneous Bandwidth (IBW)

k is the same variable used in Equation H.19

$$IBW = \frac{kc}{Lsin\theta_o}. \tag{H.20}$$

H.2.5 Grating Lobes

Grating lobe locations:

$$sin\theta_{GL} = \frac{\lambda}{\lambda_o}sin\theta_o \pm P\frac{\lambda}{d}, \quad P = 0, 1, 2, \cdots \tag{H.21}$$

Maximum element spacing d for a grating lobe free scan volume.

$$d = \frac{\lambda}{1 + sin\theta_o}. \tag{H.22}$$

H.2.6 AESA Errors

Phase shifter quantization using N bits

$$LSB = \frac{360^\circ}{2^N}. \tag{H.23}$$

Average and peak SLLs due to N bit quantization

$$AverageSLL = \frac{1}{3n_{elem}\epsilon}\frac{\pi^2}{2^{2N}},$$
$$PeakSLL = \frac{1}{2^{2N}}. \tag{H.24}$$

Average SLL due to random amplitude and phase errors

$$AverageSLL = \frac{(\pi)^{\frac{1}{2}}\overline{\epsilon^2}}{D^{\frac{1}{2}}P}. \tag{H.25}$$

ϵ^2 is the error variance, D is the directivity, and P is the probability of an element working.

H.2.7 Coordinate System Transformations

Given Antenna Angles θ_z and ϕ		
Radar coordinates	θ_{AZ}	$tan^{-1}\left(\frac{sin\theta_z \cdot cos\phi}{cos\theta_z}\right)$
	θ_{EL}	$sin^{-1}(sin\theta_z \cdot sin\phi)$
Antenna cone angles	θ_A	$sin^{-1}(sin\theta_z \cdot cos\phi)$
	θ_E	$sin^{-1}(sin\theta_z \cdot sin\phi)$

Given Radar Angles θ_{AZ} and θ_{EL}		
Antenna angles	θ_z	$cos^{-1}(cos\theta_{AZ} \cdot cos\theta_{EL})$
	ϕ	$tan^{-1}\left(\frac{sin\theta_{EL}}{sin\theta_{AZ} \cdot cos\theta_{EL}}\right)$
Antenna cone angles	θ_A	$sin^{-1}(sin\theta_{AZ} \cdot cos\,\theta_{EL})$
	θ_E	θ_{EL}

Given Antenna Cone Angles θ_A and θ_E		
Antenna angles	θ_z	$sin^{-1}\left(\sqrt{sin^2\theta_A - sin^2\theta_E}\right)$
	ϕ	$tan^{-1}\left(\frac{sin\theta_E}{sin\theta_A}\right)$
Radar angles	θ_{AZ}	$sin^{-1}\left(\frac{sin\theta_A}{sin\theta_E}\right)$
	θ_{EL}	θ_E

H.2.8 Sine Space

$$u = sin\theta_z\, cos\phi \tag{H.26}$$

$$v = sin\theta_z\, sin\phi \tag{H.27}$$

$$w = cos\theta_z \tag{H.28}$$

Conversion to Sine Space from Angular Coordinates

Sine Space	Antenna (θ_z, ϕ)	Radar $(\theta_{AZ}, \theta_{EL})$	Antenna Cone (θ_A, θ_E)
u	$sin\theta_z cos\phi$	$sin\theta_{AZ}cos\theta_{EL}$	$sin\theta_A$
v	$sin\theta_z sin\phi$	$sin\theta_{EL}$	$sin\theta_E$
w	$cos\theta_z$	$cos\theta_{AZ}cos\theta_{EL}$	$cos\left(sin^{-1}\left(\frac{sin\theta_A}{cos\theta E}\right)\right)cos\theta_E$

H.2.9 Roll, Pitch, and Yaw Formulas

$$\begin{gathered}
Roll = [R] = \begin{vmatrix} cos\theta_R & -sin\theta_R & 0 \\ sin\theta_R & cos\theta_R & 0 \\ 0 & 0 & 1 \end{vmatrix} \\
Pitch = [R] = \begin{vmatrix} 1 & 0 & 0 \\ 0 & cos\theta_P & sin\theta_P \\ 0 & -sin\theta_P & cos\theta_P \end{vmatrix} \\
Yaw = [Y] = \begin{vmatrix} cos\theta_Y & 0 & -sin\theta_Y \\ 0 & 1 & 0 \\ sin\theta_Y & 0 & cos\theta_Y \end{vmatrix}
\end{gathered} \tag{H.29}$$

H.2.10 Integrated Gain

$$G(\theta, \phi) = \frac{4\pi U(\theta, \phi)}{\int_{\phi=0}^{2\pi}\int_{\theta=0}^{\pi} U(\theta, \phi) sin\theta d\theta d\phi}. \tag{H.30}$$

U is the radiation intensity. For analysis, the AESA pattern can be calculated and then used in Equation H.30 to calculate the gain as a function of θ and ϕ. Matlab code for this is available in (Brown, 2012).

H.3 Array Elements

H.3.1 Fractional Bandwidth

$$BW_{frac} = \frac{f_{max} - f_{min}}{f_{center}}. \tag{H.31}$$

H.3.2 Polarization

Linear Polarization

$$\Delta\phi = \phi_y - \phi_x = \pm n\phi, \quad n = 0, 1, 2, \ldots \tag{H.32}$$

Tilt Angle

$$\tau = tan^{-1}\frac{E_y}{E_x} \tag{H.33}$$

Circular Polarization

$$\Delta\phi = \phi_y - \phi_x = \begin{cases} +\left(\frac{1}{2} + 2n\right)\pi, & n = 0, 1, 2, \ldots \textit{ for RHCP} \\ -\left(\frac{1}{2} + 2n\right)\pi, & n = 0, 1, 2, \ldots \textit{ for LHCP} \end{cases} \tag{H.34}$$

The orientation for RHCP is clockwise using the right-hand rule in the direction of propagation while LHCP is counterclockwise in the direction of propagation.

Elliptical Polarization

$$E_x \neq E_y, \Delta\phi = \phi_y - \phi_x = \begin{cases} +\left(\frac{1}{2} + 2n\right)\pi, & n = 0, 1, 2, \ldots \textit{ for RHEP} \\ -\left(\frac{1}{2} + 2n\right)\pi, & n = 0, 1, 2, \ldots \textit{ for LHEP} \end{cases} \tag{H.35}$$

or

$$E_x = E_y, \Delta\phi = \phi_y - \phi_x \neq \pm\frac{n}{2}\pi \begin{cases} > 0, & n = 0, 1, 2, \ldots \textit{ for RHEP} \\ < 0, & n = 0, 1, 2, \ldots \textit{ for LHEP} \end{cases} \tag{H.36}$$

RHEP and LHEP represent right-hand elliptical polarization and left-hand elliptical polarization, respectively.

Axial Ratio

$$|AR| = \frac{\textit{major axis length}}{\textit{minor axis length}} = \frac{OA}{OB} \geq 1. \tag{H.37}$$

H.3.3 Active Match

$$\Gamma_m(f, \theta_o) = \frac{V_m^-}{V_m^+} = \sum_{n=1}^{N} \frac{a_n}{a_m} S_{mn} e^{j\frac{2\pi}{\lambda}(m-n)\, d\, sin\theta_o}, \quad m = 1, 2, \ldots, M. \tag{H.38}$$

This expression is for a 1D linear array. A similar expression can be calculated for the 2D case using Equation H.17.

H.3.4 Scan Loss

Assuming a cosine expression for the element pattern with an element factor roll off, the scan loss is:

$$\begin{aligned} \textit{Scan Loss}_{dB} &= -10log_{10}(cos^{EF}(\theta)) \text{ [dB]} \\ &= -EF \cdot 10log_{10}(cos^{EF}(\theta)) \text{ [dB]} \end{aligned} \tag{H.39}$$

H.4 Transmit Receive Modules

H.4.1 Amplifier Expressions

Efficiency

$$\eta = \frac{P_{out}}{P_{DC}}. \tag{H.40}$$

Power Added Efficiency (PAE)

$$\begin{aligned} PAE &= \tfrac{P_{out}-P_{in}}{P_{DC}} \\ &= \left(1 - \tfrac{1}{G}\right)\eta. \end{aligned} \tag{H.41}$$

Output P1dB (1 dB Compression)

$$OP_{1dB} = (IP_{dB} + G) - 1 \text{ dB}. \tag{H.42}$$

Power Series Output Voltage Response

$$V_{out} = a_o + a_1 V_{in} + a_2 V_{in}^2 + a_3 V_{in}^3 + \cdots + a_n V_{in}^n. \tag{H.43}$$

nth-Order Output Intercept Point

$$OIP_n = \frac{1}{(1-n)} P_{o_n} - \frac{n}{(1-n)} P_{o_1}. \tag{H.44}$$

$n = 1$ refers to the fundamental/linear output power where $P_{o_1} = G_{dB} + P_{in}$.

H.4.2 Reliability

Element Failure Probability Using the Binomial Cumulative Distribution Function

$$P(\#\ of\ failures \leq F) = \sum_{i=0}^{F} \frac{M!}{i!(M-i)!} P^i (1-P)^{M-i} \tag{H.45}$$

P is the probability of an element failure and M is the total number of AESA elements.

Mean Time Between Failures (MTBF)

$$\begin{aligned} MTBF_{AESA} &= \tfrac{1}{\lambda_E} \sum_{i=0}^{F} \tfrac{1}{M-i} \\ &\approx \tfrac{1}{\lambda_E} \tfrac{F}{M} \\ &= MTBF_E \tfrac{F}{M} \end{aligned} \tag{H.46}$$

λ_E is the element failure rate.

Availability

$$Availability = \frac{MTBF_{AESA}}{MTBF_{AESA} + MTTR_{AESA}} \tag{H.47}$$

$MTTR$ is the mean time to repair.

H.5 Beamformer

H.5.1 General Beamformer Expressions

Number of Power Dividers in a Corporate Beamformer

$$\#of\ Power\ Dividers = log_2(M) = \frac{log_{10}(M)}{log_{10}(2)} \tag{H.48}$$

M is the total number of elements in the AESA.

Voltage Weights for a Uniform Lossless Beamformer

$$\alpha_m = \frac{1}{\sqrt{M}}. \tag{H.49}$$

Taper Loss - Beamformer Only

$$TL = \frac{1}{M} \cdot \frac{|\sum_{m=1}^{M} \alpha_m|^2}{\sum_{m=1}^{M} |\alpha_m|^2}. \tag{H.50}$$

Taper Loss - Distributed Between the TRMs and Beamformer

$$TL = \frac{1}{M} \cdot \frac{|\sum_{m=1}^{M} \beta_m \alpha_m|^2}{\sum_{m=1}^{M} |\beta_m \alpha_m|^2} \tag{H.51}$$

β_m is the TRM attenuator weights.

H.5.2 Beam Spoiling

Phase Only Weighting

$$\beta_{spoil_{m_{phase\ only}}} = e^{j\left[\left(m-\frac{(M+1)}{2}\right)\left(\frac{2\psi\sqrt{\pi}}{M-1}\right)\right]^2} \tag{H.52}$$

ψ is the beam spoil factor.

1D Pattern Using Quadratic Phase

$$F(\theta) = cos^{\frac{EF}{2}}\theta \cdot \sum_{m=1}^{M} \beta_{spoil_{m_{phase\ only}}} e^{j\left(\frac{2\pi}{\lambda}x_m sin\theta - \frac{2\pi}{\lambda_o}x_m sin\theta_o\right)}. \tag{H.53}$$

H.5.3 Monopulse AOA

3 Channel Monopulse S-Ratios

Measured amplitude ratio used to determine AOA.

$$S_{\Delta_{AZ}} = \frac{\Delta_{AZ}}{\Sigma} = -jtan(\frac{L_x}{4}k\Delta u)$$
$$S_{\Delta_{EL}} = \frac{\Delta_{EL}}{\Sigma} = +jtan(\frac{L_y}{4}k\Delta v) \tag{H.54}$$

L is the length of the aperture in the x or y dimension.

2 Channel Monopulse S-Ratio

Measured amplitude and phase used to determine AoA.

$$S_R = \frac{\Delta_R}{\Sigma} = |S_R|e^{j\Phi_R}. \tag{H.55}$$

H.6 AESA Cascaded Performance

H.6.1 Fundamental Expressions

Input Signal and Noise

$$\begin{aligned} S_{in_m} &= \frac{ERP}{4\pi R^2} \cdot A_e \cdot L_{array\ element} \\ &= \frac{ERP}{4\pi R^2} \cdot \frac{\lambda^2 D_e \cos^{EF}(\theta)}{4\pi} \cdot L_{array\ element}. \\ N_{in_m} &= kTB \cdot L_{array\ element}. \end{aligned} \tag{H.56}$$

$L_{array\ element}$ is the loss of the array element that is inclusive of radome, mismatch, and ohmic loss. D_e is the array element directivity.

Noise Factor

$$F = \frac{SNR_{in}}{SNR_{out}} = \frac{S_{in}}{S_{out}} \cdot \frac{N_{out}}{N_{in}}. \tag{H.57}$$

Nonlinear Device Output Noise Power

$$N_{out} = kTBG_aF_a. \tag{H.58}$$

Resistive Device Noise Factor

$$F = \frac{1}{\beta^2}. \tag{H.59}$$

Cascaded Gain, Noise Factor, and Noise Temperature

$$\begin{aligned} G_{a_{cascasde}} &= G_{a_1} \cdots G_{a_N}, \\ F_{cascade} &= \left(F_1 + \frac{(F_2-1)}{G_{a_1}} + \cdots + \frac{(F_N-1)}{G_{a_1} \cdots G_{a_{N-1}}}\right), \\ T_{cascade} &= TF_{cascade} \\ &= T\left(F_1 + \frac{(F_2-1)}{G_{a_1}} + \cdots + \frac{(F_N-1)}{G_{a_1} \cdots G_{a_{N-1}}}\right). \end{aligned} \tag{H.60}$$

H.6.2 AESA Cascaded Expressions

AESA Output Signal Power

$$S_{out_{AESA}} = G_{a_m} S_{in_m} \left|\sum_{m=1}^{M} \alpha_m \beta_m\right|^2. \tag{H.61}$$

AESA Output Noise Power

$$N_{out_{AESA}} = kTB \sum_{m=1}^{M} \alpha_m^2 \beta_m^2 G_{a_m} \left(F_m + \frac{\left(\frac{1}{\beta_m^2} - 1\right)}{G_{a_m}} \right). \tag{H.62}$$

AESA Noise Factor

$$F_{AESA} = \frac{\frac{S_{in}}{N_{in}}}{\frac{N_{out}}{N_{in}}} = \frac{S_{in}}{S_{out}} \cdot \frac{N_{out}}{N_{in}}$$
$$= \left(\frac{1}{M \cdot TL}\right) \left(F_m + \frac{\frac{\sum_{m=1}^{M} \alpha_m^2}{\sum_{m=1}^{M} \alpha_m^2 \beta_m^2} - 1}{G_{a_m}} \right). \tag{H.63}$$

AESA nth-Order Intercept Point

$$ip_{n_{AESA}} = \left(\frac{\left| \sum_{m=1}^{M} \sqrt{ip_{n_{amplifier}}^{(1-n)}} \alpha_m \beta_m^n \right|^2}{\left| \sum_{m=1}^{M} \alpha_m \beta_m \right|^{2n}} \right)^{\frac{1}{(1-n)}}. \tag{H.64}$$

AESA Spurious Free Dynamic Range

$$SFDR_{AESA} = \left(\frac{n-1}{n}\right) \left[\left(\frac{\left| \sum_{m=1}^{M} \sqrt{ip_{n_{amplifier}}^{(1-n)}} \alpha_m \beta_m^n \right|^2}{\left| \sum_{m=1}^{M} \alpha_m \beta_m \right|^{2n}} \right)^{\frac{1}{(1-n)}} - kTB \sum_{m=1}^{M} \alpha_m^2 \beta_m^2 G_{a_m} \left(F_m + \frac{\left(\frac{1}{\beta_m^2} - 1\right)}{G_{a_m}} \right) \right]. \tag{H.65}$$

H.7 Adaptive Beamforming

Adaptive Beamforming Weights

$$\vec{w}_{optimum} = k\vec{R}^{-1} \vec{v}(\theta_o). \tag{H.66}$$

k is an arbitrary constant that can be set to 1. It scales both the signal and noise plus interference, does not impact the SINR (signal to interference and noise ratio).

Covariance Matrix

$$\vec{R} = E\left[\vec{x}\vec{x}^H\right]$$
$$= |S|^2\vec{v}(\theta_s)\vec{v}^H(\theta_s) + \sum_{k=1}^{K} |a_k|^2\vec{v}(\theta_k)\vec{v}^H(\theta_k) + \sigma_n^2\vec{I}. \tag{H.67}$$

Steering Vector

$$\vec{v}(\theta) = \left[1 \; e^{j\frac{2\pi}{\lambda}dsin\theta} \; \dots \; e^{j\frac{2\pi}{\lambda}(M-1)dsin\theta}\right]^T. \tag{H.68}$$

Reference

Brown, A.D. *Electronically Scanned Arrays: MATLAB® Modeling and Simulation*. CRC Press, 2012.

Index

Printed and bound by CPI Group (UK) Ltd, Croydon, CR0 4YY

20/08/2026

14946426-0001